T0343417

MOLECULAR ANATOMY OF CELLULAR SYSTEMS

Progress in Biotechnology

Progress in Biotechnology 22

MOLECULAR ANATOMY
OF CELLULAR SYSTEMS

Edited by

I. Endo, T. Kudo, H. Osada, T. Shibata, I. Yamaguchi

Biochemical Systems Laboratory, Riken, Saitama University,

Hirosawa 2-1 Wako-shi, Saitama, 351-0198 Japan

2002

ELSEVIER

Amsterdam – Boston – London – New York – Oxford – Paris
San Diego – San Francisco – Singapore – Sydney – Tokyo

ELSEVIER SCIENCE B.V.
Sara Burgerhartstraat 25
P.O. Box 211, 1000 AE Amsterdam, The Netherlands

First edition 2002

Library of Congress Cataloging in Publication Data
A catalog record from the Library of Congress has been applied for.

British Library Cataloguing in Publication Data

Molecular anatomy of cellular systems. - (Progress in
   Biotechnology ; v. 22)
   1.Molecular biology 2.Cells
   I.Endo, I.
   572.8

Printed and bound in the United Kingdom
Transfered to Digital Printing, 2011
ISBN: 0-444-50739-6
ISSN: 0921-0423

⊗ The paper used in this publication meets the requirements of ANSI/NISO Z39.48-1992 (Permanence of Paper).

Cover illustration: By Dr. Yutaka Ito, Riken, Japan.

PREFACE

Design is a leading concept as well as the principal motivation for the creation of artificial systems. A successful design generally requires that the structures and functions of the elements that constitute the system as well as the principles that determine how the elements cooperate together to create function be fully understood. These requirements have not been satisfied within the fields of biotechnology and medicine. Compared to the recent emergence of artificial systems, living organisms acquired their present day structures and functions through evolution over three to four billion years. Despite the fact that the design of living organisms is recorded in the DNA sequence, our understanding of the structures and functions of the elements that constitute living organisms is very limited. To fulfill the requirements, we initiated the following approaches under our ten-year project entitled "Biodesign Research." Firstly, we tried to isolate and characterize the functional elements that constitute the organelles of various organisms. Secondly, we tried to reconstitute systems that reproduce biological functions *in vitro* from individual elements in order to understand how the elements cooperate together to yield a function. Thirdly, we attempted to resolve biological structures at various resolutions ranging from the atomic to the cellular level to further our knowledge about the fundamental principles that various functions at the molecular level and to design artificial systems.

We initiated the Biodesign Research Project in April 1990. Under this project, a number of researchers in different disciplines came together to form what became known as the Biodesign Research Group. The main group consisted of three subgroups: (a) cellular machinery, (b) membrane machinery and (c) biomolecular morphology. The long-term goal of the first group was to understand the various functions of organelles at the molecular level, in the hope of designing organelles with novel functions. The long-term goal of the second and third groups was to isolate and characterize the functional elements of organelles in various organisms, and to resolve their sub-cellular structures at various resolutions ranging from the atomic to the cellular level.

Ten years after the start of the Biodesign Research Project, although design based on the knowledge and principle has still to be realized, we are content to see that the concepts and approaches that were proposed at the outset of the project have now become widely accepted and even popular. Thus, we decided to terminate the research project in 2000, and to initiate a new one that specifically addresses the realization of biological design. In this book, we review our progress during the last ten years and outline our future plans to realize and establish the concept of design in the biological sciences.

Finally, we would like to express our sincere thanks to people from outside the group who supported the Biodesign Research Project; especially to those in the offices of RIKEN, most notably Mr. Shigekazu Saito, Mr. Hiroshi Imaizumi and Mr. Michio Seki, whose support was critical during the four year period prior to the launch of the Biodesign Research Project. We also thank Mr. Kaoru Mamiya of the former Science and Technology Agency for his encouragement during the launching of this project, and to Professor Koki Horikoshi and Dr. Tadahiko Ando for their continual advice and encouragement, which were invaluable.

November 19, 2001

Isao Endo
Isamu Yamaguchi
Toshiaki Kudo
Hiroyuki Osada
Takehiko Shibata

Contents

PART I
CELLULAR FUNCTIONS

Molecular Anatomy of Cellular Systems
I. Endo et al., (editors)

Stress meets development in p38 MAP kinase

Tatsuhiko Sudo[a][*], Masumi Maruyama[a] and Hiroyuki Osada[a]

[a] Antibiotics Laboratory, RIKEN
2-1 Hirosawa Wako Saitama 351-0198, Japan

p38 mitogen-activated protein kinases (MAPKs) form one of three major families of MAPKs and play various roles in converting extracellular stimuli into cellular responses. In the past years, extensive and intensive studies highlighted their roles in the stress responses, such as osmotic shock, UV irradiation and inflammatory cytokines. Recent genetic studies reveal their additional functions in development. This review aims to provide an overview of their functions update for better understandings of physiological and developmental roles of p38(s) and for the future development of therapeutic reagents controlling p38 signal transduction pathways.

1. INTRODUCTION

A large body of knowledge has been accumulated in the last decade regarding the signaling pathways converting extracellular stimuli through MAPKs into specific cellular responses (1-4). The MAPKs represented by extracellular signal regulated kinase (ERK), c- Jun N-terminal kinase (JNK) and p38 MAPK are one of the best characterized groups of the protein kinases and play crucial roles in the determination and execution of the cell fate leading to proliferation, differentiation and cell death. These kinases have a common feature that they are phosphorylated and activated by specific upstream kinases, MAP kinase kinase/ MAPK upstream activating kinase (MKK/MEK), and then transmit signals to the downstream target proteins by phosphorylating serine or threonine residue(s) (5-7). A prototype of the multistep phospho-relay system can be already observed in yeast and evolutionary conserved to flies and human in more complex fashions (8-11). Among them, ERK pathway is best characterized and plays a major role in converting mitogenic stimuli into cellular responses (12). JNK is a central component of the JNK pathway and was first identified as an upstream kinase for c-Jun, a major component of AP-1 transcription factors (13,14). The transcription factor as well as ATF2/CRE-BP1 is activated by phosphorylation and regulate many gene expressions (15). The newest member of the MAPK family, p38, was simultaneously

[*] Corresponding author. E-mail address: sudo@postman.riken.go.jp
Our works on p38 MAPK have been supported by Biodesign Research Program, President's Special Research Grant and Bioarchitect Research Project of RIKEN.

identified as a cytokine suppressive antiinflammatory drug (CSAID)-binding protein (16), a lipopolysaccharide (LPS)-activated kinase (17) or a stress-responsive kinase (18). p38 also plays a central role in the p38 pathway. Since latter two kinases are activated in response to the environmental stress such as UV irradiation, heat, osmotic shock, genotoxic reagents, proinflammatory cytokines and protein synthesis inhibitors, they are also categorized as stress-activated protein kinases (SAPKs). Even though each pathway has unique regulatory features, the stress stimuli often result in their activation simultaneously.

In the past years, the functional analyses of the p38 have been accelerated by the discovery and availability of the p38 specific inhibitors, such as SB202190 (16) and SB203580 (19). So far, four isoforms (α, β, γ, δ) of p38 have been cloned and characterized in mammals (16-18, 20-24), and only p38α and p38β are sensitive to these inhibitors (25). These new tools together with advanced genetic techniques, such as productions of transgenic flies and mice, facilitated functional studies of p38 not only in physiology but also in development.

This review focuses on the functions of p38 upon environmental stress and also include some of the recent genetic studies showing developmental functions of p38.

2. p38 MAP KINASE FAMILY

2.1. Members

p38α (originally named p38) was identified as a CSAID-binding protein, a LPS-activated kinase or a stress-responsive kinase. So far, four isoforms (α, β, γ, δ) of p38 have been cloned and characterized in mammals (16-18, 20-24, Table 1). They have at least 60 % homology between each of their amino acid sequences. One for yeast (8) and two for flies (26-28) had been also identified as functional homologues for p38. The identity is still high at more than 50 % between yeast high osmolarity glycerol (Hog1) gene product and human p38. Even though four members of mammalian p38s have high homology in their amino acid sequences, p38α and p38β form a subgroup based on their sensitivities to SB203580 (25). p38α and p38β show relatively broad mRNA expression profiles in human tissues (29). On the other hand p38γ and p38δ have prominent or rather restricted expression profiles in skeletal muscle, and kidney and lung respectively (20). As a common feature for p38s, they all have threonine-glycine-tyrosine (TGY) motif in their activation loops. Both threonine and tyrosine residues are phosphorylated leading to the activation of p38 (30, 31). Once p38s were phosphorylated and activated by the specific upstream kinases, then transmit these signals to the down stream targets by phosphorylating serine or threonine residue(s) followed by proline. This characteristic can be also observed in cell cycle regulated kinases, such as cdc2, to be classified as proline-directed kinases (32).

2.2. Regulations

Evolutionary conserved p38 pathways from yeast to mammals were governed by similar mechanisms. Hog1 gene product, a homologue of p38 in budding yeast, is phosphorylated and activated by PBS2 (9), the yeast homologue of MKK, in response to the high osmolarity and p38 can functionally complement Hog1 deficient yeast strain in hyperosmotic conditions (33). In mammalian cells, upstream kinases such as MKK3, MKK6 and in some cases MKK4

Table 1.
Characteristics of human p38 MAPK isoforms

Isoform	Upstream kinase	Tissue distribution	Sensitivity to SB203580
p38 (α)	MKK3, 4, 6	Broad	+
p38β	MKK6	Broad	+
p38γ	MKK3, 6	Skeletal muscle	-
p38δ	MKK3, 6	Kidney and lung	-

phosphorylate and activate p38s with different spectra. That is MKK6 phosphorylates all members of p38 MAPKs but MKK3 can't phosphorylate p38β (30). In addition to their activation by upstream kinases, the involvement of p62, first identified as a phosphorylation independent ligand of p56^{lck}, in the regulation of p38 activity with a substrate dependent manner *in vitro* is recently proposed (34). The novel mechanism of the regulation of p38 activity somehow helps to clarify how various environmental stimuli exert different responses in cell type and/or in stimulation dependent fashions by using a limited number of signal transducers.

2.3. Targets
To date, many downstream targets for p38 have been described. These include transcription factors, such as ATF2 (18), MEF2C (35), CHOP (36), C/EBPβ (37), c-Jun (38) and so on. As mentioned above, c-Jun is phosphorylated in the activation domain by JNK (13). The phosphorylated residue(s) by p38 seemed to be different from the ones by JNK, suggesting the organized regulations of its transcriptional activity by different members of the closely related kinases (38). Other than transcription factors, cytoplasmic proteins represented by MAPKAPK2/3 (39, 40), MNK1/2 (41, 42) and MBP are also phosphorylated by p38. Among them, a functional relevance of the phosphorylation *in vivo* by p38 is well characterized with MAPKAPK2. MAPKAPK2 is phosphorylated and activated by p38 to phosphorylate HSP27 to regulate actin dynamics in cells (43). Moreover MAPKAPK2 is shown to be involved in the p38 regulation through controlling subcellular localizations (44).

2.4. Specific inhibitors
Recent discovery of p38 MAPKs specific inhibitors accelerate the functional analyses both *in vivo* and *in vitro*. Several p38 specific inhibitors are currently used in various approaches for our understandings of p38s. A series of pyridinyl imidazole compounds, represented by SB202190 and SB 203580, is originally identified by their abilities to suppress the production of inflammatory cytokines (Fig. 1). Then, this class of compounds has shown to specifically inhibit p38α and p38β activities by binding to their ATP pockets but not those of p38γ nor

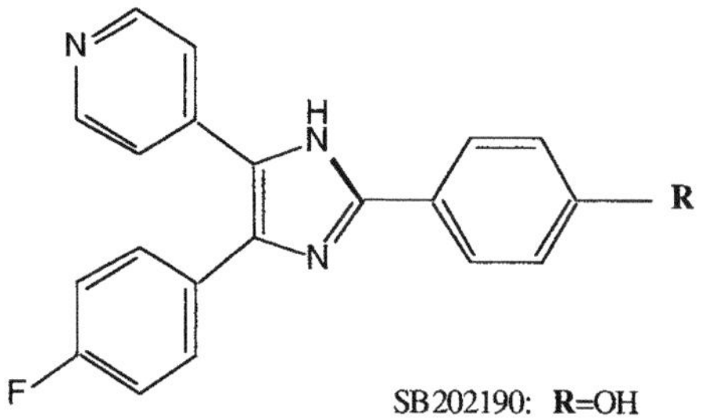

SB 202190: **R=OH**
SB 203580: **R=SOCH$_3$**

Figure 1. Chemical structures of p38 MAPK inhibitors

p38δ (25). This specificity is explained by the amino acid alignments at the back of ATP pockets, namely amino acid 106 to 108 (Thr-His-Leu) confer their strong competitivebinding affinities only for p38α and p38β, but not for p38γ and p38δ (45-47). The developments of specific inhibitors and/or activators for different isoforms of p38 will be the key toward the next step to assign their functions precisely.

3. FUNCTIONS

3.1. Inflammation

The roles of the p38 in inflammation is well documented both *in vitro* and *in vivo*. Direct evidences, such as anti-inflammatory drugs bind to p38 and productions of pro-inflammatory cytokines are reduced by the inhibition of p38 kinase activity, strongly suggest their crucial roles in the inflammatory responses. Also the observations that the expressions of many inflammatory related proteins, including COX-2 and iNOS, were reduced by the treatment with p38 inhibitors support their roles (48, 49).

3.2. Cell fates

The activation of the p38 MAPKs leads to various consequences in cultured cells with cell type dependent manners. Activation of p38 is generally accepted to lead apoptosis (50-52). But in the case of Jurkat T cells, even after stimulation with Fas leading to apoptosis accompanied by p38 activation, the activation itself is not necessary for the consequence (53).

Though p38 pathway has a negative role in cell cycle progression at either G1-S (54) or M phase in NIH 3T3 cells and at M phase in *Xenopus* cell free system (55), it is required for the proliferation of Swiss 3T3 cells induced by FGF-2 but not by serum (56). These results indicate that p38 pathway is involved in cell proliferation differently.

Myogenic differentiation from C2C12 myoblasts to myotubes is induced by the activation of p38 pathway (57). Several studies also provide supportive evidences with other cell lines showing the involvement of the myocyte enhancer factor 2C (MEF2C) and MyoD in the p38 pathway leading to myogenic differentiation (58-60). Adipocytic differentiation was also

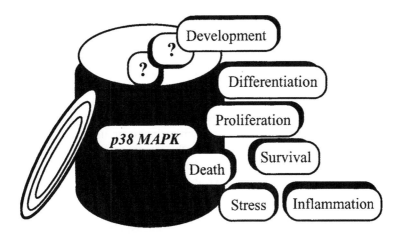

Figure 2. Key words around p38 MAPK

blocked by p38 specific inhibitors (37) and was induced by the activation of p38 pathway with the ectopically expressed constitutive active form of MKK6 (61). These studies also suggested that C/EBPβ might be responsible for the adipogenic differentiation through p38 pathway. Neuronal differentiation again appears to be mediated by p38 pathway in PC12 cells (62, 63). A recent study also shows that the p38 pathway is involved in the osteoclastogenesis mediated by receptor-activation of NF-κB ligand (64).

3.3.Development

Recent advances in the genetic approaches facilitated the analyses of the developmental functions of the specific genes in *Drosophila*. Especially with respect to MAPK pathways, three major MAPK pathways are well conserved but less complicated, since smaller numbers of the molecules are involved in these pathways (65). Transgenic flies expressing the dominant negative form of D-p38b, one of the *Drosophila* homologues of p38, in the wing showed similar phenotypes of decapentaplegic (dpp:TGF-β homologue) and thick vein (tvk:dpp receptor) mutant. D-p38b may be involved in the wing development by transducing signals from dpp to ATF-2 via TAK-1 (28, 66-68).

So far four reports based on the observation of mice lacking in p38α gene locus have shown that p38α is essential for the mouse development (69-72). The results were somehow surprising, since this is the first example that mice lacking only one isoform of MAPK family died before birth. This may reflect functional redundancy of the isoforms of ERK and JNK but not p38 during the mouse development. It is also possible that there is the same functional redundancy but the expression of p38β is too low to complement p38α deficiency in mice. It is just beginning to understand physiological and developmental functions of p38. Further assignments of their functions to each isoform(s) not only in development but also in physiological conditions will be promoted by the productions of the mice lacking different isoforms and/or conditionally knocked out these genes.

4. AN UNSOLVED MYSTERY AND FUTURE PROSPECTS

Here comes a simple question. How can p38 play such a numerous roles in the physiological conditions? In spite of the fact that p38s have only been appeared in less than a decade, tremendous efforts have made possible to show their participations in many physiological events so quickly (Fig. 2). It is obvious that the discovery and availability of p38 specific inhibitors have made a great contribution to the research of p38 pathway and others. As any inhibitor is not absolutely specific, improper usage or overdose of p38 inhibitors may mislead the conclusions. Moreover the ectopic expression of active or inactive forms of kinases in the p38 pathway may affect neighboring pathways resulting in the incorrect interpretation of the phenomena. We should keep in mind that such troubles can be evaded by the well designed experiments with proper controls as well as by right interpretations from different approaches.

One of the great contributions of these findings mentioned above to our society is to pull the trigger on the search for new therapeutic targets in p38 pathway for inflammatory treatments. For the next step, it is also hopeful to find new drugs can be applied in the treatments for diseases, yet unknown, caused by p38 deficiency.

REFERENCES

1. C. S. Hill and R. Triesman, Cell, 80 (1995) 199.
2. M. Cobb and E. J. Goldsmith, J. Biol. Chem., 270 (1995) 14843.
3. M. Karin, J. Biol. Chem., 270 (1995) 16483.
4. C. J. Marshall, Cell, 80 (1995) 179.
5. R. J. Davis, Trends Biochem. Sci., 19 (1994) 470.
6. K. J. Blumer and G. L. Johnson, Trends Biochem. Sci., 19 (1994) 236.
7. J. M. Kyriakis and J. Avruch, J. Biol. Chem., 271 (1996) 24313.
8. J. L. Brewster, T. de Valoir, N. D. Dwyer, E. Winter and M. C. Gustin, Science, 259 (1993) 1760.
9. T. Maeda, M. Takekawa and H. Saito, Science, 269 (1995) 554.
10. S. J. Han, K.Y. Choi, P.T. Brey and W.J. Lee, J. Biol. Chem., 273 (1998) 369.
11. E. Cano and L. C. Mahadevan, Trends Biochem. Sci., 20 (1995) 117.
12. M. J. Robinson and M. H. Cobb, Curr. Opin. Cell Biol., 9 (1997) 180.
13. M. Hibi, A. Lin, T. Smeal, A. Minden and M. Karin, Genes Dev., 7 (1993) 2135.
14. B. Derijard, M. Hibi, I. H. Wu, T. Barrett, B. Su. T. Deng, M. Karin and R. J. Davis, Cell, 76 (1994) 1025.
15. S. Gupta, D. Campbell, B. Derijard and R. J. Davis, Science, 267 (1995) 389.
16. J. C. Lee, J. T. Laydon, P. C. McDonnell, T. F. Gallagher, S. Kumar, D. Green, D. McNulty, M. J. Blumenthal, J. R. Heys, S. W. Landvatter, J. E. Strickler, M. M. McLaughlin, I. R. Siemens, S. M. Fisher, G. P. Livi, J. R. White, J. L. Adams and P. R. Young., Nature, 372 (1994) 739.
17. J. Han, J. D. Lee, L. Bibbs and R. J. Ulevitch, Science, 265 (1994) 808.
18. J. Rouse, P. Cohen, S. Trigon, M. Morange, A. Alonso-Llamazares, D. Zamanillo, T. Hunt and A. R. Nebreda, Cell, 78 (1994) 1027.

19. A. Cuenda, J. Rouse, Y. N. Doza, R. Meier, P. Cohen, T. F. Ballagher, P. R. Young and J. C. Lee, FEBS Lett., 364 (1995) 229.
20. Y. Jiang, H. Gram, M. Zhao, L. New, J. Gu, L. Feng, P. F. Di, R. J. Ulevitch and J. Han, J. Biol. Chem., 272 (1997) 30122.
21. B. Stein, M. X. Yang, D. B. Young, R. Janknecht, T. Hunter, B. W. Murray and M. S. Barbosa, J. Biol. Chem., 272 (1997) 19509.
22. Z. Li. Y. Jiang, R. J. Ulevitch and J. Han, Biochem. Biophys. Res. Commun., 228 (1996) 334.
23. M. Goedert, A. Cuenda, M. Craxton, R. Jakes and P. Cohen, EMBO J., 17 (1997) 3563.
24. X. S. Wang,, K. Diener, C. L. Manthey, S. Wang. B. Rosenzweig, J. Bray, J. Delancy, C. N. Cole, H. P. Chan, N. Mantlo, H. S. Lichenstein, M. Zukowski and Z. Yao, J. Biol. Chem., 272 (1997) 23668.
25. S. Kumar, M. S. Jiang, J. L. Adams and J. C. Lee, Biochem. Biophys. Res. Commun., 263 (1999) 825.
26. S. J. Han, K. Y. Choi, P. T. Brey and W. J. Lee, J. Biol. Chem., 273 (1998) 369.
27. Z. S. Han, H. Enslen, X. Hu, X. Meng, I. H. Wu, T Barrett, R. J. Davis and Y. T. Ip, Mol. Cell. Biol., 18 (1998) 3527.
28. T. Adachi-Yamada, M. Nakamura, K. Irie, Y. Tomoyasu, Y. Sano, E Mori, S. Goto, N. Ueno, Y. Nishida and K . Matsumoto, Mol. Cell. Biol., 19 (1999) 2322
29. Y. Jiang, C. Chen, Z. Li, W. Guo, J. A. Gegner, S. Lin and J. Han, J. Biol. Chem., 271 (1996) 17920.
30. B. Derijard, J. Raingeaud, T. Barrett, I. Wu, J. Han, R. J. Ulevitch and R. J. Davis, Science, 267 (1995) 682.
31. H. Enslen, J. Raingeaud, R. and R. J. Davis, J. Biol. Chem., 273 (1998) 1741.
32. U. Marklund, G. Brattsand, V. Shingler and M. Gullberg, J. Biol. Chem., 268 (1993) 15039.
33. S. Kumar, M. M. McLaughlin, P. C. McDonnell, J. C. Lee, G. P. Livi and P. R. Young, J. Biol. Chem., 270 (1995) 19043.
34. T. Sudo, M. Maruyama and H. Osada, Biochem. Biophys. Res. Commun., 269 (2000) 521.
35. J. Han, Y. Jiang, V. Li. V. Kravchenko and R. J. Ulevitch, Nature, 386 (1997) 296.
36. X. Z. Wang and D. Ron, Science, 2732 (1996) 1347.
37. J. A. Engelman, M. P. Lisanti and P. E. Scherer, J. Biol. Chem., 273 (1998) 32111.
38. T. Sudo and M. Karin, Meth. Enzymol., 322 (2000) 88.
39. D. Stokoe, D. G. Campbell, S. Nakielny, H. Hidaka, S. J. Leevers, C. Marshall and P. Cohen, EMBO J., 11 (1992) 3985.
40. M. M. McLaughlin, S. Kumar, P. C. McDonnell, S. Van Horn, J. C. Lee, G. P. Livi and P. R. Young, J. Biol. Chem., 271 (1996) 8488.
41. A. J. Waskiewicz, A. Flynn, C. G. Proud and J. A. Cooper, EMBO J., 16 (1997) 1909.
42. R. Fukunaga and T. Hunter, EMBO J., 16 (1997) 1921.
43. J. Guay, H. Lambert, G. Gingras-Breton, J. N. Lavoie, J. Huot and J. Landry, J. Cell Sci., 110 (1997) 357.
44. R. Ben-levy, S. Hooper, R. Wilson, H. F. Paterson and C. J. Marshall, Curr. Biol., 8 (1998) 1049.
45. P. R. Young, M. M. McLaughlin, S. Kumar, S. Kassis, M. L. Doyle, D. McNulty, T. F.

Gallagher, S. Fisher, P. C. McDonnell, S. A. Carr, M. J. Huddleston, G. Seibel, T. G. Porter, G. P. Livi, J. L. Adams and J. C. Lee, J. Biol. Chem., 272 (1997) 12116.

46. B. Frantz, T. Klatt, M. Pang, J. Parsons, A. Rolando, H. williams, M. J. Tocci, S. J. O'Keefe and E. A. O'Neill, Biochemistry, 37 (1998) 13846.

47. J. M. Lisnock, A. Tebben, B. Frantz, E. A. O'Neill, G. Croft, S. J. O'Keefe, Li, C. Hacker, S. de Laszlo, A. Smith, B. Libby, N. Liverton, J. Hermes and P. LoGrasso, Biochemistry, 37 (1998) 16573.

48. T. Shalom-Barak, J. Quanch and M. Lotz, J. Biol. Chem., 273 (1998) 27467.

49. K. Subbaramaiah, W. J. Chung and A. J. Dannenberg, J. Biol. Chem., 273 (1998) 32943.

50. Z. Xia, M Dickens, J. Raingeaud, R. J. Davis and M. E. Greenberg, Science, 270 (1995) 1326.

51. P. Schwenger, P. Bellosta, I. Vietor, C. Basilico, E. Y. Skolnik and J. Vilcek, Proc. Natl. Acad. Sci. USA, 94 (1997) 2869.

52. H. Kawasaki, T. Morooka, S. Shimohama, J. Kimura, T. Hirano, Y. Gotoh and E Nishida, J. Biol. Chem., 272 (1997) 18518.

53. S. Huang, Y. Jiang, Z. Li, E. Nishida, P. Mathias, S. Lin, R. J. Ulevitch, G. R. Nemerow and J. Han, Immunity, 6 (1997) 739.

54. A. Molnar, A. M. Theodoras, L. I. Zon and J. M. Kyriakis, J. Biol. Chem., 272 (1997) 13229.

55. K. Takenaka, T. Moriguchi and E. Nishida, Science, 280 (1998) 599.

56. P. Mapher, J. Biol. Chem., 274 (1999) 17491.

57. A. Cuenda and P. Cohen, J. Biol. Chem., 274 (1999) 4341.

58. A. Zetser, E. Gredinger and E. Bengal, J. Biol. Chem., 274 (1999) 5193.

59. P. L. Puri, Z. Wu, P.Zhang, L. D. Wood, K. S. Bhakta, J. Han, J. R. Feramisco, M. Karin and J. Y. J. Wang, Genes Dev., 14 (2000) 574.

60. Z. Wu, P. J. Woodring, K. S. Bhakta, K. Tamura, F. Wen, J. R. Feramisco, M. Karin, J. Y. J. Wang and P. L. Puri, Mol. Cell. Biol., 20 (2000) 3951.

61. J. A. Engelman, A. H. Berg, R. Y. Lewis, A. Lin, M. P. Lisanti and P. E. Scherer, J Biol. Chem., 274 (1999) 35630.

62. T. Morooka and E. Nishida, J. Biol. Chem., 273 (1998) 24285.

63. S. Iwasaki, M. Iguchi, K Watanabe, R. Hoshino, M. Tsujimoto and M. Kohno, J. Biol. Chem., 274 (1999) 26503.

64. M. Matsumoto, T. Sudo, T. Saito, H. Osada and M. Tsujimoto, J. Biol. Chem., 275 (2000) 31155.

65. E. Martin-Blanco, Bioessay, 22 (2000) 637.

66. N. Perrimon, Curr. Opin. Cell Biol., 6 (1994) 260.

67. J. Riese, X. Yu, A. Munnerlyn, S. Eresh, S. C. Hsu, R. Grosschedl, M. Bienz, Cell, 88 (1997) 777.

68. K. Yamaguchi, K. Shirakabe, H. Shibuya, K Irie, I. Oishi, N. Ueno, T. Taniguchi, E. Nishida and K. Matsumoto, Science, 270 (1995) 2008.

69. M. Allen, L. Svensson, M. Roach, J. Hamber, J. McNeish and C. A. Gabel, J. Exp. Med., 191 (2000) 859.

70. R. H. Adams, A. Porras, G. Alonso, M. Jones, K. Vintersten, S. Panelli, A. Valladares, S. Panelli, A. Valladares, L. Perez, R. Klein and A. R. Nebreda, Mol. Cell, 6 (2000) 109.

71. K. Tamura, T. Sudo, U. Senftleben, A. M. Dadak, R. Johnson and M. Karin, Cell, 102

(2000) 221.
72. J. S. Mudgett, J. Ding, L. Guh-siesel, N. A. Chartrain, L. Yang, S. Gopal and M. M. Shen, Proc. Natl. Acad. Sci. USA, 97 (2000) 10454.

Molecular Anatomy of Cellular Systems
I. Endo et al., (editors)

Molecular dissection of cytotoxic functions mediated by T cells

Takao Kataoka and Kazuo Nagai

Department of Bioengineering, Tokyo Institute of Technology
4259 Nagatsuta-cho, Midori-ku, Yokohama 226-8501, Japan

Cytotoxic functions mediated by T cells play a central role in host defense and regulation of the immune system. In response to antigen stimulation, T cells produce various cytokines that transmit positive or negative signals, and directly kill harmful cells such as virus-infected cells. The perforin/granzyme system and the Fas/Fas ligand system are primarily involved in the killing mechanisms mediated by T cells. To clarify the molecular basis of T cell-mediated cytotoxicity, we took advantage of specific inhibitors as biochemical tools. This article describes the identification and characterization of useful "bioprobes" that dissect cytotoxic functions mediated by T cells.

1. INTRODUCTION

Cytotoxic T lymphocytes (CTLs) utilize various cytotoxic molecules to eliminate unwanted cells such as virus-infected and transformed cells [1]. CD8[+] CTLs kill target cells primarily by the perforin/granzyme system and Fas ligand, whereas CD4[+] CTLs mainly exert Fas ligand-dependent cytotoxicity. The pore-forming protein perforin and the serine protease granzyme B are enriched in specialized granules known as lytic granules [2]. CTLs conjugate tightly with target cells upon activation via T cell receptors (TCRs) in concert with additional adhesion molecules, and are triggered to release the lytic granules toward the interface between CTLs and target cells. Granzyme B enters cells via the cation-independent mannose 6-phosphate/insulin-like growth factor receptor [3], and perforin facilitates the nuclear and/or cytosolic translocation of granzyme B [4,5]. Granzyme B cleaves and activates various key substrates essential for apoptosis induction [6,7]. Moreover, CTLs express membrane-bound forms of Fas ligand on the cell surface upon antigen stimulation, and Fas ligand cross-links Fas receptors on the target cells which induces self-processing of the initiator caspase, procaspase-8 [8]. Activated caspase-8 cleaves downstream substrates such as caspase-3, committing cells to apoptosis. Although many essential molecules involved in CTL-mediated cytotoxicity are currently known, the molecular mechanism linking TCR activation and functional expression of cytotoxicity still remains to be elucidated. To gain an insight into the molecular basis of CTL-mediated cytotoxicity, we took a strategy to employ specific inhibitors as biochemical tools. In this article, we present useful "bioprobes" that dissect perforin- and Fas-based CTL-mediated cytotoxicity.

2. BIOPROBES FOR T CELL-MEDIATED CYTOTOXICITY

To measure perforin-dependent target cell lysis, the alloantigen-specific CD8$^+$ CTL clone OE4 was preincubated with test samples for 2 h, and then cultured with ^{3}H-TdR- or ^{51}Cr-labeled Fas-negative (or low expressing) target cells for 4 h. Cytotoxic activity was measured by the release of radioactivity into culture supernatants. We applied this assay system to known compounds, microbial metabolites, and plant extracts, and succeeded to identify several inhibitors. These compounds are categorized into three groups (Figure 1):(i) bioprobes that inhibit vacuolar acidification [9,10], (e.g. concanamycin A (CMA), bafilomycin A$_1$ (BMA), prodigiosin 25-C (PRG 25-C), destruxin E (DRE)), (ii)bioprobes that

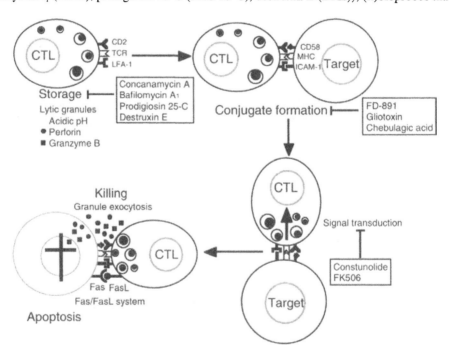

Figure 1. Classification of specific inhibitors that block CTL-mediated cytotoxicity. Lytic granules are acidic organelles where perforin and granzyme B are enriched. CTLs conjugate with target cells via the binding of TCRs to peptides in the context of MHC molecules, together with the interaction of LFA-1 (lymphocyte function-associated antigen-1) and CD2 (LFA-2) with ICAM-1 (intercellular adhesion molecule-1) and CD58 (LFA-3), respectively. TCR engagement leads to activation of several protein tyrosine kinases and phosphorylation of many signaling molecules. Following intracellular events involve Ca^{2+} mobilization, activation of serine/threonine kinases and initiation of signaling pathways such as the Ras-dependent MAPKs (mitogen-activated protein kinases) cascade. These biochemical changes trigger to release the lytic granules (granule exocytosis) and induce the expression of membrane-bound Fas ligand. Granzyme B and Fas induce apoptosis in the target cells. The sites of action of various inhibitors are represented.

block conjugate formation, (e.g. FD-891 [11], gliotoxin [12], chebulagic acid [13]), and (iii) bioprobes that block early signal transduction via TCR (e.g. costunolide [14], FK506 [15]).

3. ACIDIFICATION IS ESSENTIAL TO MAINTAIN THE STRUCTURE AND FUNCTION OF LYTIC GRANULES

3.1. Inactivation and proteolytic degradation of perforin in lytic granules upon neutralization of acidic pH by concanamycin A

CMA belongs to 18-membered macrolide antibiotics. CMA specifically inhibits vacuolar type H^+-ATPase and thereby prevents the functions of acidic organelles [16]. We found that CMA markedly inhibits perforin-dependent target cell lysis mediated by OE4 cells [9]. The profound blockage of the cytolytic activity was only observed when OE4 cells, but not target cells, were treated with CMA [9]. However, CMA did not block conjugate formation between CTLs and target cells [9]. Thus, it seemed that CMA does not affect early signaling events downstream of T cell activation. CMA prevented the acidification of vacuolar organelles [9], and raised the internal pH of the lytic granules from 5.7 to around neutral [17]. Although CMA marginally impaired granule exocytosis in response to CD3 stimulation, perforin activity in the lytic granules was completely abrogated by CMA treatment. In agreement with this finding, CMA markedly reduced the cellular level of perforin, while a significant amount of perforin was still detectable in OE4 cells treated with CMA [9]. Since the serine protease inhibitor diisopropylfluorophosphoridate (DFP) antagonized the perforin decrease in CMA-treated cells [18], perforin seems to undergo proteolytic degradation by serine proteases. Nevertheless, residual unhydrolyzed perforin is likely to be inactivated upon neutralization of acidic pH in the lytic granules, as DFP did not reverse perforin activity in CMA-treated cells [18]. Similarly to CMA, perforin inactivation and perforin degradation were observed in OE4 cells treated with BMA [10], a 16-membered macrolide antibiotic that inhibits vacuolar type H^+-ATPase [16]. In the cell-free systems using perforin-enriched granule fractions, perforin activity was completely suppressed by acidic pH [18]. Moreover, acidic pH rendered perforin more resistant to calcium-induced inactivation [18]. In the lytic granules, perforin associates with chondroitin sulfate A at acidic pH, and perforin binding to chondroitin sulfate A is dependent on pH [19]. Thus, it was proposed that acidic pH plays a role in the efficient packaging of perforin keeping it quiescent. Figure 2 depicts a model for the molecular events in the lytic granules upon neutralization of acidic pH. The CMA-induced increase of intragranular pH might initially trigger the dissociation of perforin from chondroitin sulfate A. Resultant free forms of perforin are subjected to be inactivated in a calcium-dependent manner and subsequently hydrolyzed by DFP-sensitive serine proteases (possibly certain granzymes). To keep lytic machinery functional, acidic pH plays an essential role to maintain the integrity of perforin within the lytic granules.

3.2. Inactivation of perforin in lytic granules upon neutralization of acidic pH by prodigiosin 25-C and destruxin E

PRG 25-C has three conjugate pyrrole rings together with a long hydrophobic hydrocarbon chain, and is known as a red pigment produced by *Streptomyces* and *Serratia*. We have recently demonstrated that prodigiosins are a new group of H^+/Cl^- symporters that uncouple proton translocators [20,21,22]. In intact cells, PRG 25-C uncouples vacuolar type H^+-ATPase, inhibits vacuolar acidification, and affects glycoprotein processing [23]. The drastic swelling of Golgi apparatus and mitochondria was also observed in PRG 25-C-treated cells [23]. Structure-activity relationship revealed that three pyrrole units are essential for the inhibitory activity of prodigiosins toward vacuolar acidification [24]. DRE belongs to a group of cyclic depsipeptides, and was reported to inhibit endosomal acidification [25]. In addition, a close structural analogue of DRE, destruxin B was shown to prevent vacuolar type H^+-ATPase [26]. We found that PRG 25-C and DRE markedly block perforin-dependent killing pathway [10]. However, in contrast to CMA and BMA, PRG 25-C and DRE failed to reduce the cellular amount of perforin, although these agents markedly reduced perforin activity in the lytic granules [10]. These observations suggest that perforin is dominantly inactivated even without proteolysis in cells treated with PRG 25-C and DRE. This might be explained by the notion that PRG 25-C and DRE do not increase the intragranular pH to the same level as CMA and BMA, and that perforin inactivation might occur at lower pH than that required for the proteolytic degradation of perforin.

3.3.Morphological changes of lytic granules induced by concanamycin A

We observed that CMA induces drastic morphological changes of the lytic granules in OE4 cells [9]. Electron microscopic observations revealed that the lytic granules present in OE4 cells are morphologically heterogeneous, yet commonly consisted of homogeneously stained cores and numerous small vesicles on the peripheral regions [9]. In CMA-treated OE4 cells, the peripheral small vesicles were dramatically decreased, and the structure of the cores became rough and loose [9]. Empty vacuoles were frequently seen in CMA-treated OE4 cells [9,17]. Thus, these observations demonstrate that acidification is essential to maintain the structure of the lytic granules. The perforin-deficient $CD8^+$ CTL clone P0K and the $CD4^+$ CTL clone BK-1 that lacks the expression of perforin harbored cytoplasmic granules devoid of cores but filled with numerous small vesicles [27]. Thus, perforin might not be required for the biogenesis of entire granules but be essential for the formation of the cores.

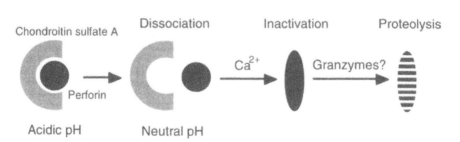

Figure 2. Inactivation and proteolytic degradation of perforin in lytic granules upon neutralization of acidic pH.

4. PERFORIN- AND FAS LIGAND-DEPENDENT KILLING PATHWAYS UTILISE DISTINCT SECRETORY ROUTES UPON ACTIVATION

We screened specific agents that selectively block either perforin-dependent or Fas ligand-dependent killing pathway [28]. Perforin-dependent killing activity was measured by cytolysis of Fas-low expressing mastocytoma P815 cells by the CD8⁺ CTL clone OE4, whereas Fas ligand-dependent killing activity was measured by cytolysis of Fas-expressing B cell line A20 cells by the CD4⁺ CTL clone BK-1. Perforin-dependent cytotoxicity mediated by OE4 cells was markedly blocked by CMA and BMA, whereas it was insensitive to actinomycin D (an inhibitor of RNA synthesis), cycloheximide (an inhibitor of protein synthesis), brefeldin A (BFA; an inhibitor of protein transport), and tunicamycin (an inhibitor of glycosylation) [28]. By contrast, Fas ligand-dependent killing pathway was inhibited by actinomycin D, cycloheximide, BFA, and tunicamycin, but totally insensitive to CMA and BMA [28]. Thus, Fas ligand-based cytotoxicity depends on *de novo* RNA/protein synthesis, glycosylation and the intracellular protein transport through Golgi apparatus. By contrast, perforin-based cytotoxicity does not require *de novo* RNA/protein synthesis, but depends on the acidification of intracellular organelles, suggesting that a sufficient amount of perforin is already stored in the lytic granules for target cell lysis before TCR activation. Different inhibitor sensitivity clearly indicates that perforin- and Fas ligand-dependent killing pathways utilize distinct secretory machineries upon TCR activation (Figure 3).

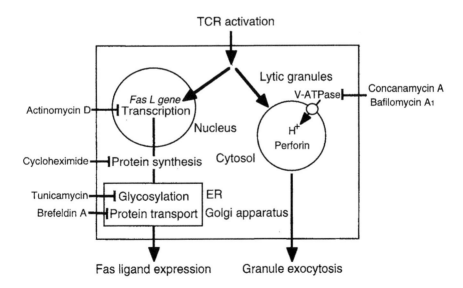

Figure 3. Distinct secretory routes utilized by perforin- and Fas ligand-dependent killing pathways upon TCR activation. Concanamycin A and bafilomicin A₁ inhibit vacuolar H⁺-ATPase (V-ATPase), and rise the internal pH of the lytic granules. Neutralization of acidic pH induces inactivation and proteolysis of perforin within the lytic granules.

5. CONCANAMYCIN A IS A SPECIFIC INHIBITOR TO BLOCK PERFORIN-MEDIATED KILLING PATHWAY

The observation that CMA markedly inhibits perforin-dependent cytolysis but does not affect Fas ligand-dependent cytolysis mediated by CD4$^+$ and CD8$^+$ CTL clones prompted us to generalize that CMA is an inhibitor that blocks perforin-based, but not Fas ligand-based, CTL-mediated cytotoxicity. We found that CMA completely blocks perforin-dependent killing pathway by a wide range of bulk CTLs at 10-100 nM [27]. However, Fas ligand-based CTL-mediated killing pathway was hardly influenced even at higher concentrations of CMA [27]. Thus, our results demonstrate that CMA is a powerful tool to block only perforin-dependent CTL-mediated cytotoxicity, and to clarify the contribution of two distinct cytolytic pathways. By contrast, BFA markedly inhibited Fas ligand-based cytotoxicity, whereas perforin-based cytotoxicity was only marginally perturbed by BFA [27]. Quantitative comparison of CMA-sensitive and insensitive cytolytic activity revealed that CD8$^+$ CTLs mainly exert perforin-based killing pathway and complementary Fas ligand-based killing pathway.

6. AN ESSENTIAL ROLE OF VACUOLAR TYPE H$^+$-ATPASE IN THE SURVIVAL OF MATURE CD8$^+$ CTLs

6.1. Prodigiosin 25-C specifically blocks the induction of CD8$^+$ CTLs in allogeneic tumor-immunized mice

PRG 25-C and metacycloprodigiosin were initially isolated from *Streptomyces hiroshimensis* as active substances that preferentially inhibit concanavalin A (Con A)-induced T cell proliferation rather than lipopolysaccharide-induced B cell proliferation [29]. When PRG 25-C was injected into mice immunized with allogeneic tumors and sheep red blood cells (SRBC), PRG 25-C selectively suppressed the induction of CTL activity, although PRG 25-C did not affect the production of anti-SRBC antibody [30,31]. Neither IL-2 production nor IL-2 receptor expression were significantly impaired by PRG 25-C [31]. These findings suggest that PRG 25-C is an immunosuppressant that is specific for CTL induction *in vivo*, and that its mode of action is distinct from those of other immunosuppressive drugs such as FK506 and cyclosporin A. Although an inhibitor of vacuolar type H$^+$-ATPase concanamycin B (CMB) is structurally unrelated to PRG 25-C, CMB profoundly suppressed the killing activity of CTLs in mice immunized with allogeneic tumors [32]. Moreover, PRG 25-C and CMB markedly reduced the increase of CD8$^+$ cells without affecting the number of CD4$^+$ cells and B cells [32]. These observations indicate that the immunosuppressive activity of these agents toward the induction of CTLs results from the blockage of vacuolar acidification. The suppression of CTL activity and the decrease in activated CD8$^+$ cells were obvious even after the single injection of PRG 25-C and CMB when allogeneic CD8$^+$ CTLs are fully induced [33]. Thus, mature CD8$^+$ CTLs seem to be the direct target of PRG 25-C and CMB *in vivo*.

6.2. Vacuolar type H⁺-ATPase activity might be required for the survival of CD8⁺ CTLs

CMA preferentially reduced the cell viability of CD8⁺ cells prepared from immunized mice, compared with those of CD4⁺ and B220⁺ populations [34]. The CD8⁺ CTL clone, but not the CD4⁺ T cell clones, underwent apoptosis when treated with CMA over 8 h [34]. Anti-CD3 stimulation or phorbol ester accelerated the reduction of cell viability only in the CD8⁺ CTL clone [34]. However, this rapid cell death was not accompanied by DNA fragmentation and nuclear condensation [34]. These results suggest that vacuolar type H⁺-ATPase activity is critical for the survival of mature CD8⁺ CTLs. Although it was reported that CMA induces apoptosis in several cell lines [35,36], CD8⁺ CTLs seem to be most sensitive to CMA among the cells so far reported. The high susceptibility of CD8⁺ CTLs to CMA might be caused by the leakage of cytotoxic molecules such as granzyme B from the lytic granules into the cytoplasm, since CMA induced the drastic morphological changes in the lytic granules [9]. By contrast, the rapid cell death of CD8⁺ CTLs induced by CMA and phorbol ester (or anti-CD3) was independent of perforin and granzyme B [34], but might be dependent on cytosolic pH, since intracellular acidification proceeds caspase activation during apoptosis [37,38], and the up-regulation of vacuolar type H⁺-ATPase that pumps protons out of the cytosol delays the onset of apoptosis [39]. Inasmuch as CD8⁺ CTLs harbor many lytic granules, these cells are likely to contain a large number of vacuolar type H⁺-ATPase. Thus, we propose that vacuolar type H⁺-ATPase is involved in the regulation of the cytosolic pH in CD8⁺ CTLs and plays an essential role in the survival of CD8⁺ CTLs especially when activated.

7. PRODIGIOSIN 25-C IS A NOVEL IMMUNOSUPPRESSANT THAT BLOCKS T CELL ACTIVATION

Con A-induced proliferation of murine T cells was selectively inhibited by PRG 25-C [29,31]. T cell proliferation was increasingly suppressed by PRG 25-C in the presence of higher amount of Con A regardless of T cell subpopulations [40]. The enhancement of the suppressive effects of PRG 25-C was not observed when murine T cells were activated with phytohemagglutinin, anti-CD3 antibody, or allogeneic antigen presenting cells [40]. Although various tumor cell lines exhibited increased sensitivity to PRG 25-C in the presence of Con A, mouse T lymphoma YAC-1 cells became most susceptible to PRG 25-C when cultured with Con A [41]. Since PRG 25-C significantly up-regulated Con A-binding sites on the cell surface possibly through affecting glycoprotein processing [23,41], the increased binding of Con A by PRG 25-C might result in the augmentation of Con A toxicity. Recently, we synthesized a series of PRG analogues, and observed that a prodigiosin analogue inactive to block vacuolar acidification still displays Con A-dependent suppression of T cell proliferation [24]. Inhibitors of vacuolar type H⁺-ATPase and polyether ionophores, both of which perturb vacuolar acidification, suppressed YAC-1 proliferation more strongly in the presence of Con A, albeit inefficiently compared with PRG 25-C [41]. Thus, it is possible that PRG 25-C converts Con A-induced proliferative signals to death signals through a mechanism independent of the blockage of vacuolar acidification. It was reported that PRG 25-C blocks the activation of human T cells in mid to late G1, and inhibits the induction of several cell cycle-associated cyclins and kinases as well as the phosphorylation of retinoblastoma proteins [42]. The PRG analogue PNU156804 was demonstrated to efficiently block IL-2-

dependent proliferation and the activation of the transcription factors NF-κB and AP-1 [43]. Therefore, it seems likely that PRG 25-C targets signaling molecule(s) in early T cell activation.

8. BIOPROBES THAT BLOCK CONJUGATE FORMATION

8.1. FD-891

FD-891 belongs to 18-membered macrolide antibiotics, and structurally related to CMA. FD-891 was originally isolated from the fermentation broth of *Streptomyces graminofaciens* A-8890 as an anti-tumor agent [44,45]. We found that FD-891 markedly blocked both perforin- and Fas ligand-based CTL-mediated cytotoxicity [11]. However, FD-891 failed to affect vacuolar acidification and perforin activity in the lytic granules [11]. Nevertheless, FD-891 markedly blocked conjugate formation between CTLs and target cells when CTLs are treated with FD-891 [11]. FD-891 significantly reduced the expression of the TCR/CD3 complex, but not other adhesion molecules such as LFA-1 [11]. Thus, the blockage of conjugate formation might be partly due to the FD-891-induced down-regulation of the TCR/CD3 complex.

8.2. Gliotoxin

Gliotoxin is a potential etiologic agent that is synthesized by pathogenic fungi such as *Aspergillus fumigatus*, and exhibits a variety of immunosuppressive activities such as the blockage of NF-κB activation [46]. We found that gliotoxin profoundly inhibits both perforin- and Fas ligand-dependent killing pathways mediated by CTLs [12]. Gliotoxin markedly blocked CTL binding to immobilized anti-CD3 antibody as well as conjugate formation by targeting CTLs [12]. Nevertheless, gliotoxin unaffected the expression of surface molecules such as TCR and CD3. We postulate that gliotoxin impairs affinity and/or accumulation of surface proteins or cytoskeletal organization that supports efficient conjugate formation.

8.3. Chebulagic acid

Chebulagic acid and gallic acid were isolated from the extract of a herbal medicine, kashi (myrobalans: the fruit of *Terminalia chebula*) as active compounds that block CTL-mediated cytotoxicity [13]. Granule exocytosis in response to anti-CD3 stimulation was blocked by gallic acid and chebulagic acid [13]. CTL binding to immobilized anti-CD3 antibody was significantly reduced by chebulagic acid (data not shown), suggesting that chebulagic acid might prevent conjugate formation.

9. BIOPROBES THAT BLOCK EARLY SIGNAL TRANSDUCTION VIA T CELL RECEPTORS

9.1. Costunolide

Costunolide and dehydrocostus lactone were purified from the extract of a herbal medicine, mokko (*Saussurea lappa* Clarke) as active principles that block CTL-mediated cytotoxicity [14]. The structure-activity relationship revealed that α-methylene-γ-butyrolactone structures that undergo Michael reaction with SH-containing components are essential for the inhibitory effects [14,47]. Tyrosine phosphorylation in response to CD3 stimulation was significantly inhibited by costunolide at the concentration that did not affect CTL binding to immobilized anti-CD3 antibody [14]. Thus, costunolide might bind to cysteine residues of tyrosine kinases, and prevent an increase in tyrosine phosphorylation that is a critical step for T cell activation.

9.2.FK506

FK506 is a potent immunosuppressant, and blocks the production of lymphokines such as IL-2 in T cells via the blockage of nuclear translocation of the transcription factor NF-AT [48,49]. FK506 binds to FK506-binding proteins (FKBPs), and the FKBP-FK506 complex inhibits the calcium- and calmodulin-dependent serine/threonine phosphatase calcineurin that is required for the nuclear translocation of NF-AT [48,49]. We found that FK506 partially blocks perforin-dependent target cell lysis mediated by a $CD8^+$ CTL clone [15]. In agreement with the previous report [50], the release of granzyme A and B was prevented by FK506, although a significant granzyme release was still detectable even in the presence of excess FK506 [15]. By contrast, IFN-γ production was completely abrogated by FK506 [15]. Granule exocytosis was inefficiently induced by the combination of ionomycin and phorbol ester capable of mimicking TCR activation to trigger IFN-γ production [15]. Granule exocytosis and IFN-γ production in response to TCR stimulation were profoundly prevented by an inhibitor of protein kinase C, calphostin C [15]. Thus, these results suggest that granule exocytosis depends on the activation of protein kinase C, and requires either calcineurin-dependent or independent additional signals downstream of TCR activation.

10. CONCLUSIONS

Cytotoxic functions mediated by T cells play a central role in host defense and regulation of the immune system. Malfunction of the killing mechanisms often causes autoimmune diseases and tumorigenesis. Perforin and Fas ligand are two major molecules involved in the killing mechanisms mediated by T cells. Different inhibitor sensitivity demonstrates that perforin- and Fas ligand-based killing pathways utilize distinct secretory machineries upon TCR activation. Molecular dissection with use of vacuolar type H^+-ATPase inhibitors clearly reveals that acidification via vacuolar type H^+-ATPase plays a critical role to maintain the structure of the lytic granules as well as to keep the integrity of perforin. Further studies based on specific bioprobes will give us novel regulatory mechanisms and/or target molecules valuable to design therapeutic strategies.

REFERENCES

1. S. Shresta, C. T. Pham, D. A. Thomas, T. A. Graubert and T. J. Ley, Curr. Opin. Immunol., 10 (1998) 581.
2. L. J. Page, A. J. Darmon, R. Uellner and G. M. Griffiths, Biochim. Biophys. Acta, 1401

(1998) 146.

3. B. Motyka, G. Korbutt, M. J. Pinkoski, J. A. Heibein, A. Caputo, M. Hobman, M. Barry, I. Shostak, T. Sawchuk, C. F. B. Holmes, J. Gauldie and R. C. Bleackley, Cell, 103 (2000) 491.

4. C. J. Froelich, K. Orth, J. Turbov, P. Seth, R. Gottlieb, B. Babior, G. M. Shah, R. C. Bleackley, V. M. Dixit and W. Hanna, J. Biol. Chem., 271 (1996) 29073.

5. L. Shi, S. Mai, S. Israels, K. Browne, J. A. Trapani and A. H. Greenberg, J. Exp. Med., 185 (1997) 855.

6. R. V. Talanian, X. Yang, J. Turbov, P. Seth, T. Ghayur, C. A. Casiano, K. Orth and C. J. Froelich, J. Exp. Med., 186 (1997) 1323.

7. F. Andrade, S. Roy, D. Nicholson, N. Thornberry, A. Rosen and L. Casciola-Rosen, Immunity, 8 (1998) 451.

8. A. Ashkenazi and V. M. Dixit, Science, 281 (1998) 1305.

9. T. Kataoka, K. Takaku, J. Magae, N. Shinohara, H. Takayama, S. Kondo and K. Nagai, J. Immunol., 153 (1994) 3938.

10. K. Togashi, T. Kataoka and K. Nagai, Immunol. Lett., 55 (1997) 139.

11. T. Kataoka, A. Yamada, M. Bando, T. Honma, K. Mizoue and K. Nagai, Immunol., 100 (2000) 170.

12. A. Yamada, T. Kataoka and K. Nagai, Immunol. Lett., 71 (2000) 27.

13. S. Hamada, T. Kataoka, J.-T. Woo, A. Yamada, T. Yoshida, T. Nishimura, N. Otake and K. Nagai, Biol. Pharm. Bull., 20 (1997) 1017.

14. M. Taniguchi, T. Kataoka, H. Suzuki, M. Uramoto, M. Ando, K. Arao, J. Magae, T. Nishimura, N. Otake and K. Nagai, Biosci. Biotech. Biochem., 59 (1995) 2064.

15. T. Kataoka and K. Nagai, Immunol. Lett., 72 (2000) 49.

16. S. Dröse and K. Altendorf, J. Exp. Biol., 200 (1997) 1.

17. T. Kataoka, M. Sato, S. Kondo and K. Nagai, Biosci. Biotech. Biochem., 60 (1996) 1729.

18. T. Kataoka, K. Togashi, H. Takayama, K. Takaku and K. Nagai, Immunol., 91 (1997) 493.

19. D. Masson, P. J. Peters, H. J. Geuze, J. Borst and J. Tschopp, Biochemistry, 29 (1990) 11229.

20. T. Sato, H. Konno, Y. Tanaka, T. Kataoka, K. Nagai, H. H. Wasserman and S. Ohkuma, J. Biol. Chem., 273 (1998) 21455.

21. H. Konno, H. Matsuya, M. Okamoto, T. Sato, Y. Tanaka, K. Yokoyama, T. Kataoka, K. Nagai, H. H. Wasserman and S. Ohkuma, J. Biochem., 124 (1998) 547.

22. S. Ohkuma, T. Sato, M. Okamoto, H. Matsuya, K. Arai, T. Kataoka, K. Nagai and H. H. Wasserman, Biochem. J., 334 (1998) 731.

23. T. Kataoka, M. Muroi, S. Ohkuma, T. Waritani, J. Magae, A. Takatsuki, S. Kondo, M. Yamasaki and K. Nagai, FEBS Lett., 359 (1995) 53.

24. A. Fürstner, J. Grabowski, C. W. Lehmann, T. Kataoka and K. Nagai, ChemBioChem., 2 (2001) 60.

25. S. Naganuma, N. Kuzuya, K. Sakai, K. Hasumi and A. Endo, Biochim. Biophys. Acta, 1126 (1992) 41.

26. M. Muroi, N. Shiragami and A. Takatsuki, Biochem. Biophys. Res. Commun., 205 (1994) 1358.

27. T. Kataoka, N. Shinohara, H. Takayama, K. Takaku, S. Kondo, S. Yonehara and K. Nagai,

J. Immunol., 156 (1996) 3678.

28. T. Kataoka, M. Taniguchi, A. Yamada, H. Suzuki, S. Hamada, J. Magae and K. Nagai, Biosci. Biotech. Biochem., 60 (1996) 1726.
29. A. Nakamura, K. Nagai, K. Ando and G. Tamura, J. Antibiot., 39 (1986) 1155.
30. A. Nakamura, J. Magae, R. F. Tsuji, M. Yamasaki and K. Nagai, Transplantation, 47 (1989) 1013.
31. R. F. Tsuji, M. Yamamoto, A. Nakamura, T. Kataoka, J. Magae, K. Nagai and M. Yamasaki, J. Antibiot., 43 (1990) 1293.
32. M.-H. Lee, T. Kataoka, J. Magae and K. Nagai, Biosci. Biotech. Biochem., 59 (1995) 1417.
33. M.-H. Lee, T. Kataoka, N. Honjo, J. Magae and K. Nagai, Immunol., 99 (2000) 243.
34. K. Togashi, T. Kataoka and K. Nagai, Cytotechnology, 25 (1997) 127.
35. T. Nishihara, S. Akifusa, T. Koseki, S. Kato, M. Muro and N. Hanada, Biochem. Biophys. Res. Commun., 212 (1995) 255.
36. S. Akifusa, M. Ohguchi, T. Koseki, K. Nara, I. Semba, K. Yamato, N. Okahashi, R. Merino, G. Nunez, N. Hanada, T. Takehara and T. Nishihara, Exp. Cell Res., 238 (1998) 82.
37. R. A. Gottlieb, J. Nordberg, E. Skowronski and B. M. Babior, Proc. Natl. Acad. Sci. USA, 93 (1996) 654.
38. S. Matsuyama, J. Llopis, Q. L. Deveraux, R. Y. Tsien and J. C. Reed, Nat. Cell Biol., 2 (2000) 318.
39. R. A. Gottlieb, H. A. Giesing, J. Y. Zhu, R. L. Engler and B. M. Babior, Proc. Natl. Acad. Sci. USA, 92 (1995) 5965.
40. T. Kataoka, J. Magae, H. Nariuchi, M. Yamasaki and K. Nagai, J. Antibiot., 45 (1992) 1303.
41. T. Kataoka, J. Magae, K. Kasamo, H. Yamanishi, A. Endo, M. Yamasaki and K. Nagai, J. Antibiot., 45 (1992) 1618.
42. S. Songia, A. Mortellaro, S. Taverna, C. Fornasiero, E. A. Scheiber, E. Erba, F. Colotta, A. Mantovani, A.-M. Isetta and J. Golay, J. Immunol., 158 (1997) 3987.
43. A. Mortellaro, S. Songia, P. Gnocchi, M. Ferrari, C. Fornasiero, R. D'Alessio, A. Isetta, F. Colotta and J. Golay, J. Immunol., 162 (1999) 7102.
44. M. Seki-Asano, T. Okazaki, M. Yamagishi, N. Sakai, K. Hanada and K. Mizoue, J. Antibiot., 47 (1994) 1226.
45. M. Seki-Asano, Y. Tsuchida, K. Hanada and K. Mizoue, J. Antibiot., 47 (1994) 1234.
46. H. L. Pahl, B. Krauss, K. Schulze-Osthoff, T. Decker, E. B. Traenckner, M. Vogt, C. Myers, T. Parks, P. Warring, A. Muhlbacher, A. P. Czernilofsky and P. A. Baeuerle, J. Exp. Med., 183 (1996) 1829.
47. S. Yuuya, H. Hagiwara, T. Suzuki, M. Ando, A. Yamada, K. Suda, T. Kataoka and K. Nagai, J. Nat. Prod., 62 (1999) 22.
48. S. L. Schreiber and G. R. Crabtree, Immunol. Today, 13 (1992) 136.
49. A. Rao, Immunol. Today, 15 (1994) 274.
50. J. P. Dutz, D. A. Fruman, S. J. Burakoff and B. E. Bierer, J. Immunol., 150 (1993) 2591.

Molecular Anatomy of Cellular Systems
I. Endo et al., (editors)

Molecular imaging of the cytoskeleton using GFP-actin fluorescence microscopy

Yuling Yan[a] and Gerard Marriott[b]

[a] Department of Medical Physics, University of Wisconsin,
1530 Medical Science, Madison, WI 53706
[b] Department of Physiology, University of Wisconsin,
1300 University Avenue, Madison, WI 53706

1. Introduction

The actin cytoskeleton is a complex network of polarized filaments that is involved in many essential processes including motility and cytokinesis, tumor cell transformation [2] and metastasis [3,9]. Actin filaments bind to at least 60 different actin-binding proteins (ABPs) in the cell that can regulate their assembly and disassembly, their organization and rheological properties [14,19]. The actin cytoskeleton, as a dynamic network, constantly undergoes cycles of assembly and disassembly. In the lamellipodium of a motile cell for example, actin filaments are assembled at their (+)-end close to the plasma membrane and the same filament network is partially disassembled a few microns behind the membrane at their (-)-ends. On the other hand actin filaments in the cell cortex turnover on a longer timescale presumably because these filaments are stabilized by extensive crosslinking. The organization and dynamics of actin filaments therefore differ according to their location in the cell – while the mechanisms regulating these properties are not fully understood, they probably involves localized activation of ABPs by molecular signals generated in receptor activated signaling pathways [13,24,25]. Therefore to understand the role of actin in cell migration, cytokinesis and transformation we need quantitative techniques that can map changes in the cellular distribution and activities of the actin cytoskeleton with high spatial and temporal resolution. Yu-Li Wang and D. Lansing Taylor were the first to demonstrate that these goals could be achieved using fluorescence microscope based imaging of fluorescent actin conjugates microinjected into living cells [4,11,18]. While this approach works well for imaging studies conducted over a few hours, it is not suitable for extended observations because the fluorescent protein conjugate is proteolyzed within the cell leading to a progressive deterioration in image quality [4]. This factor limits the usefulness of the technique in studying the role of the actin cytoskeleton in differentiation, wound healing or metastasis, which often take in excess of 24 hours. Perhaps the best chance to overcome these limitations is to image fusion proteins of actin and green fluorescent protein (GFP, [12]) since the GFP-actin gene is continuously expressed in transfected cells while any proteolysis will generate a non-fluorescent fragment. Although GFP-actin has been reported to be a suitable probe of the actin cytoskeleton in yeast, Dictyostelium and in certain mammalian cells [6,7,22,23] it is important to recognize that GFP-actin does not always behave as the endogenous actin – in particular it can inhibit the contraction of actomyosin leading to dominant negative effects [22]. On the other hand when used at tracer levels in the cell it faithfully reports on changes in the ortganization and the dynamics of the actin cytoskeleton and can do so over an extended time. In this article we show how GFP-actin can be used to quantify properties of actin in lamellipodia of motile cells, in the propulsion of Listeria bacteria through infected cells and to detail changes in the organization of actin filaments in the epithelia to mesenchyme transformation (EMT) of rat bladder NBT-II carcinoma [2,20] over a 24-hour period. Understanding the mechanism that regulates actin filament assembly and disassembly in these processes may lead to new approaches to halt the spread of tumor cell migration or to increase the rate of wound-healing. This review article summarizes our past experience in using GFP-actin as probe of the organization and dynamics of the actin cytoskeleton [22,23].

The Dictyostelium actin-15 gene, which shares considerable sequence identity with mouse β-actin, is introduced into the EcoRI cloning site of the pEGFP-C2 vector using standard cloning methods. This cloning strategy introduces a relatively hydrophilic 14 amino acid peptide linker, **SGRTQISSSSFEFK**, derived from the multiple cloning sites in the plasmid, between GFP and Dd actin. NBT-II, PC12, NIH 3T3 clone 7 and Swiss 3T3 cell lines are transfected with the GFP-actin plasmid, or the pEGFP-C2 plasmid, which serves as a control, using Lipofectin (Gibco) or calcium phosphate [5]. Transfected cells expressing GFP-actin, or GFP, are established in DMEM/ 10% FCS containing 1 mg/ml of geneticin over a period of 3-4 weeks [1]. Hela cells are transfected with GFP- actin and imaged the following day.

Expression levels of GFP-actin in permanently transfected cells are quantified using Western blot analysis using affinity-purified antibodies raised against GFP or Dd actin. Total cell lysates of non-and NGF-differentiated PC12 cells are probed with these antibodies according to the figure legend (figure 1). This figure shows that GFP-actin (A, B, lanes 3-6) is expressed at similar levels in pre- and post BGF-treated PC12 cells where it accounts for 5% of the endogenous actin. The control GFP gene (B, lanes 1 and 2) is expressed to a similar amount as GFP-actin. The amount of GFP-actin is therefore similar to a level comparable to that used with microinjected fluorescent actin [4]. Western blot analysis is an important aspect of the GFP-actin imaging approach yet researchers rarely make efforts to conduct these experiments. The need for these studies is highlighted by the finding that prolonged growth of GFP-actin transfected cells often leads to the expression of GFP-gene fusions having different localization properties compared to the full-length gene fusion.

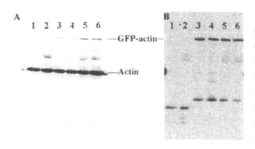

Figure 1. Western blot analysis of GFP-actin in stable transfected PC12 cells. Blots were probed with a mouse monoclonal antibody against *Dictyostelium* actin (A), and rabbit polyclonal antibodies against GFP (B). Cell lysates of non-differentiated (lanes 1, 3 and 4) and NGF differentiated cells (lanes 2, 5 and 6). Lanes 1 and 2 are lysates from cells expressing the GFPgene and lanes 3-6 were cells expressing GFP-actin gene

These clones probably express partially deleted GFP-actin genes, which no longer integrate into the endogenous actin. Using freshly thawed GFP-actin transfected cells and never culturing these cells for more than six weeks may overcome this problem. The 32 kD GFP-stained band seen in the blot (Fig. 1B, lanes 3-6) probably represents a proteolyzed fragment of GFP-actin generated during cell lysis, since the ratio of GFP-actin to the 32 kD band is different in the four PC12 samples.

The suitability of GFP-actin as a probe of the actin cytoskeleton in living cells is evaluated on the basis of three different criteria. First, GFP-actin should fully incorporate into the cytoskeleton of diverse mammalian cell lines. Second, images of GFP-actin fluorescence should respond to changes in the organization and dynamics of actin filaments such as in the lamellipodia and in the actin tail that propels Listeria bacteria in infected cells. Third, GFP-actin should not interfere with cell functions that depend on the actin cytoskeleton. We could not distinguish between fluorescence images of GFP-actin derived from either Dictyostelium actin or mouse β-actin. This is perhaps not too surprising since Dictyostelium actin and non-muscle β-actin differ by only 19 highly conservative amino acid exchanges. Both actin sub-types exhibit similar binding preferences for actin-binding proteins. While the actin subtype does not influence the function of actin in cells, the linker between GFP and the N-terminus of actin is critical. GFP-actin made with a 5 amino acid linker does not translocate on myosin II if the GFP-actin content in the filament exceeds 30% [22].

2. Cell Culture

Cells harboring the GFP-actin gene are maintained as permanent lines in DMEM/10% FCS containing 1 mg/ml G_{418} in an incubator at 37°C, 90% humidity with 10% CO_2. Cell culture is conducted in Costar or Corning petri-dishes without polylysine or coated with collagen (Sigma, St. Louis, Mo, USA). The methods and techniques used to culture and infect mammalian cells with Listeria are described in Sechi et al [16]. For microscope-based observations, cells are plated on cover slips (24 mm diameter) that had been washed with 0.1 N HCl and coated with 20 μg/ml polylysine (30,000 MW), or collagen at 0.1 mg/ml.

3. Microscopy

For microscope based observations, cells grown on glass cover-slips are mounted in a temperature controlled chamber at 36-37°C (Medical Systems Inc, N.Y.) or else in a holder on a microscope stage enclosed within a Perspex housing maintained at 37°C with a thermostat. A HEPES based DMEM containing 10% FCS is used for experiments that required long-term observations of cells on the microscope stage. Excitation of GFP-actin in cells with blue-green light is achieved with a 100 W mercury lamp using a HQ fluorescein filter (Chroma Technologies, Brattleboro, VT). The excitation filter selects the blue-green emission lines of the mercury-arc lamp while the emission is collected between 510-540 nm. Cy3 and Cy3.5 emission images are collected using custom-made filter sets. Images are recorded using a Photometrics NU200 and collected in binning mode (2 x 2) for 0.2-1 second. Image analysis and processing is made using various software including AVS (Digital),

Adobe Photoshop (3.1), NIH Image (5.2) and Coral Draw (7.0). High contrast fluorescence images of cells in growth medium are recorded using a Zeiss 100x, quartz fluar objective (NA 1.3), which is also employed in phase-contrast imaging. Confocal fluorescence microscopy of GFP-actin is performed using an instrument described in Westphal et al [22].

GFP-actin fusion proteins used in our group fulfill the three citerea as a functional probe of the cytoskeleton outlined above.

1. Images of the distribution and dynamics of GFP-actin in various mammalian cells are recorded with custom modified microscope that is optimized for GFP fluorescence detection [8, 22]. Representative fluorescence images of GFP-actin in NIH 3T3 cells are shown in figure 2 (A, B). The brilliant GFP-actin fluorescence in the lamellipodia at the leading edge

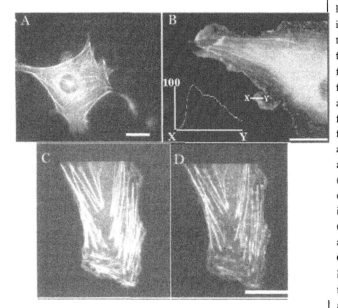

Figure 2 (A-D). Fluorescence microscopy images that show the expression and localization of GFP-actin in living stably transfected NIH 3T3 fibroblasts.). (A), GFP-actin labeling of thickened stress fibers in a living NIH 3T3 cell – notice the probe is completely absent from the nucleus. (B), GFP-actin fluorescence labels a dense actin filament mesh in the lamellipodia as well as micro-spikes and filopodia. In addition the probe faithfully reports the sites of actin in the cortex, focal contacts and stress fibers. The inset figure (B) shows how the exponential decay of GFP-actin fluorescence intensity within the lamellipodia (between the points marked X and Y). (C,D), GFP-actin and Cy3.5-phalloidin respectively label the same structures: fluorescence images of GFP-actin in Swiss 3T3 cells permeabilized with 0.2% Triton-X100 and simultaneously fixed with 0.25% glutaraldehyde. The actin cytoskeleton in these cells was also visualized using Cy3.5-phallodin. The bar shown on the lower right hand side represents 10 μm.

of the motile NIH 3T3 cell decays from the plasma membrane (sites labeled X and Y in figure 2B). This is a characteristic property of rapid actin filament turnover [17]. These images of GFP-actin also show specific labeling of filopodia and microspikes within lamellipodia, focal contacts, a thin border of cortical actin and actin stress fibers. As expected GFP-actin is completely absent from the nucleus. Images of cells expressing the GFP gene alone do not show any cytoskeletal staining. The intrinsic fluorescence of control fibroblast cells transfected with an empty plasmid is far weaker fluorescence than GFP-actin or GFP labeled cells and is restricted to small vesicles [23].

Images of GFP-actin and Cy3.5-phalloidin, a new probe for F-actin, in fixed NIH 3T3 cells are almost identical - coincident probe labeling is seen in stress fibers and lamellipodia. The images of GFP-actin and Cy3.5-phalloidin of cells shown in figure 2 (C,D) show that both probes label the same focal contacts and stress fibers. The fluorescence intensity of GFP-actin in cells co-labeled with saturating amounts of Cy3-phalloidin is extremely weak. This is a result of GFP-actin and Cy3-phalloidin exhibiting ideal photophysical properties for fluorescence resonance energy transfer (FRET). Since energy transfer occurs when the distance between GFP and Cy3 is within 5 nm [8], this observation proves that the two probes label identical actin filaments. FRET significantly decreases the quantum yield of GFP fluorescence while sensitizing Cy3-phalloidin fluorescence (data not shown). The probability of FRET in double labeling experiments can be significantly reduced using GFP-actin and a sub-stoichiometric amount of Cy3-phalloidin [8].

2. The ability of GFP-actin acting as a probe of actin filament dynamics is evident from quantitative and qualitative image analysis of changes in the distribution and intensity of GFP-actin fluorescence within lamellipodia of motile cells (Fig. 2B) and in Hela cells transfected with GFP-actin and then infected with Listeria (Fig. 3A, B). GFP-actin localizes

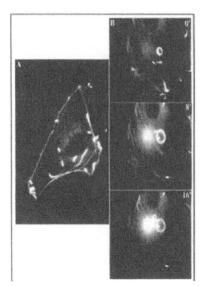

Figure 3. (A), Distribution of GFP-actin in a live Hela cell transiently transfected with the GFP-actin vector and infected with *Listeria monocytogenes*. (B) Three frames from an image sequence show the position of a motile bacterium and the evolution, and decay, of its GFP-actin filamentous tail as a function of time. The white bar shown on the lower right hand side of figure 3A represents 10 μm.

in stress fibers of HeLa cells shown in figure 3 (A,B), and in actin comet tails that form at the base of motile Listeria. Fluorescence images of GFP-actin shown in Fig. 3B, show the bacteriumis propelled in a circular fashion by actively polymerizing actin at its base. The GFP-actin comet tails decays exponentially from the base of the bacterium with a lifetime on the order of 3 minutes, which is consistent with other studies using extrinsically labeled actin microinjected into living cells [14,19]. Interestingly the bacteria are non-motile during cell division – at this stage of the cell cycle GFP-actin is uniformly distributed on the Listeria surface (Fig. 3A).

4. GFP-actin does not interfere with cell processes that require a functional cytoskeleton

Images of GFP-actin fluorescence in PC12 cells show uniform labeling in the cell body, with a slight enrichment at the cortex. No stress fibers are present in these cells. Addition of nerve growth factor to these cells induces differentiation that is accompanied by the growth of axon-like structures from the cell body (Fig. 4A, C). Images of GFP-actin fluorescence recorded at an early (Fig. 4A, B), or later stage (Fig. 4 C, D) in the differentiation process show an accumulation of actin in dynamic, needle-like filopodia in extending growth cones and these are indistinguishable from those present in control cells [15].

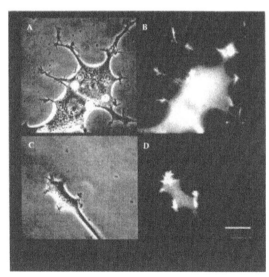

Figure 4. Live cell phase contrast (A, C), and GFP-actin fluorescence (B, D), images of PC12 cells transfected with the GFP-actin vector and treated with nerve growth factor after 24 hours (A, B), and 24 hours (C, D). The growth cones at the tips of the neurite-like extensions contain filopodia, which appear as spikes in the phase contrast images, and were densely packed with GFP-actin. The fluorescence signal in the cell body lacked any obvious structure apart from a slight enrichment at the cortex. The look-up table was adjusted in order to show the structure of filopodia in the growth cones. The white bar shown on the lower right hand side of figure 4D represents 10 μm.

5. Photo-bleaching properties of GFP-actin in live cells

Singlet oxygen radicals generated during the photo-bleaching of fluorescent dyes is a serious problem that limits the usefulness of fluorescence microscopy in studying molecular process in living cells. Singlet oxygen can inactivate proteins within a few nanometers of the photobleached fluorophore [10,21]. In the case of fluorescein-labeled tubulin, these radicals sever microtubules. As Gregorio Weber would put it, the fluorescein probe no longer performs as a spectator in the physiological drama but an actor. In addition the secondary products of singlet oxygen that often lead to a hazy fluorescence background that diminishes image quality. A particularly attractive property of GFP probe is that any radicals produced by the fluorophore will be sequestered within the GFP protein matrix, which is about 3 nm in diameter. This idea is also supported by image data showing that the cytoskeleton of GFP-actin transfected cells observed after a 2 second or 200 second irradiation are, aside from the intensity difference, almost identical (Fig. 5 A, B). Unlike extrinsically labeled fluorescent probes, which remain in the cell even after proteolysis of their host protein, GFP-actin fluorescence is lost upon proteolysis, a property that facilitates long term imaging studies.

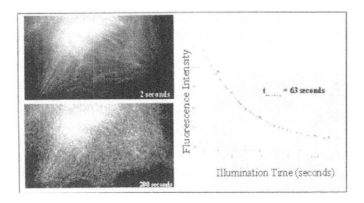

Figure 5: Photobleaching of GFP-actin in a live NIH 3T3 cell using the blue-green excitation lines of the 100-W mercury-arc selected with the FITC-Hi Q excitation filter (Chromotech, Brattleboro, VT). Fluorescence image of the cell recorded (A), after 2 seconds of irradiation, and (B), after 200 seconds of irradiation. (C), Data analysis of the GFP-actin intensity in a single region of the cell as a function of the irradiation time is analyzed as an single exponential and has a decay time of 63 seconds.

6. Changes in the organization of the actin cytoskeleton during the EMT of NBT-II cells

Since GFP-actin undergoes constant cycles of synthesis and degradation to non-fluorescent products, it is an ideal probe for long-term imaging studies of the cytoskeleton. A dramatic change in the organization of actin filaments is seen in rat NBT-II carcinoma cells during their growth factor induced EMT [2]. Cobblestone-like epithelial colonies of NBT-II cells are maintained in part by Adherens junctions, and supported by an actin ring at the cell cortex (Fig. 6A; [20]). Fluorescence images of GFP-actin in epithelia cells show that the probe is coincident with Adherens junctions. Treating epithelia cells with 100 ng/ml of EGF triggers an EMT from which emerge highly motile mesenchyme cells [2,20]. This transformation process, which is complete within 24 hours, is categorized by three distinct cell morphologies distinguishable by the distribution of GFP-actin fluorescence.

1. In epithelia cells GFP-actin is enriched at the cell cortex, at sites of cell-cell contact while it is generally uniform in the cell body (Fig. 6A,B; [2]). Immunofluorescence studies were used to show that the distribution of _-catenin in epithelia cells was coincident with Cy3.5-phalloidin at adherens junctions [23]. No obvious stress fibers were detected in these cells from images of GFP-actin in live cells, or Cy3.5-phalloidin in fixed cells [23]. About 12-16 hours after the growth factor treatment a second population of cells was found that was more evident at the periphery of a colony and appeared to be a hybrid between the epithelia and mesenchyme cells. These cells were loosely associated within a colony but not through adherens junctions, since the _-catenin label was absent from the membrane (data not shown). A thick ring of actin cables surrounds the cell cortex in the apical region of the cell (Fig. 6D). At a later stage (16-24 hours) the lamella increases in area and exhibits a bright GFP-actin fluorescence.

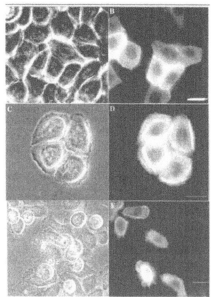

Figure 6: Phase-contrast (A, C and E), and GFP-actin (B, D and F), images of live NBT-II epithelia cells recorded before (A and B), after 12 hours (C and D), and 24 hours (E and F) of the addition of 40 ng/ml aFGF (bar 20 μm). The white bar shown on the lower right hand side of the images represents 10 μm.

A third freely motile cell population appears within 20-24 hours of EGF treatment 6E,F; ref 2]. The fluorescence of GFP-actin in these cells is generally uniform in the broad lamella except for focal contacts that sometimes give rise to stress fibers. Higher levels of GFP-actin appear at sites of membrane ruffling at the leading edge of motile cells.

7. Summary

Fluorescence imaging of GFP-actin empowers the researcher with a unique tool to image changes in the organization and dynamics of the actin cytoskeleton in living cells over an extended period of time (>24 hours). While high levels of GFP-actin expression are deleterious, lower levels found in permanently transfected cell lines (5%) do not impair cell structure and function nor does this level alter dynamic properties of the actin cytoskeleton seen in motility and cytokinesis. The continuous expression and degradation of GFP-actin in cells allows the researcher to conduct fluorescence microscope based investigations of the cytoskeleton in single living cells for many days. GFP-actin is a faithful reporter of the actin cytoskeleton and it was shown to localize to stress fibers, focal adhesion contacts and in the cell cortex in a manner indistinguishable from endogenous actin in a variety of cell types. GFP-actin can be used in a quantitative manner to probe parameters underlying the regulation of actin filament turnover. In particular it is enriched in protruding lamellipodia, membrane ruffles, filopodia and in the comet tails that form behind motile Listeria bacteria. The decay of GFP-actin fluorescence signal in these dynamic regions decreases exponentially from the site of filament growth (Fig. 2A, 3A, [14,17]. The continuous expression of the fusion protein in certain cells, together with the high photostability of the GFP-fluorophore, the lack of oxygen radical-mediated damage all contribute to the suitability of the probe to study long-term changes of the actin cytoskeleton during the EMT of NBT-II cells. Furthermore, the sensitivity of the fusion protein to the dynamics of actin filament assembly and disassembly makes it a suitable probe to image the regulation and role of actin polymerization/depolymerization in the protrusion of the leading edge and ultimately cell motility.

REFERENCES

1. Boom, R., C.J. Sol, R.P. Minnaar, J.L. Geelen, A.K. Raap, and J. van der Noordaa, J. Gen. Vir., 69 (1988) 1179.
2. Boyer, B., G.C. Tucker, A.M. Valles, W.W. Franke, and J.P. Thiery, J. Cell Biol., 109 (1989) 1495.
3. Button, E., C. Shapland and D. Lawson, J. Cell Motil. Cytoskel., 30 (1995) 247.
4. Conrad, P.A., M.A. Nederlof, I.M. Herman, and D.L. Taylor, Cell Motility & the Cytoskeleton, 14 (1989) 527.
5. Davis, L.G., M.D. Dibnar, and J.F. Battey. in Basic Methods in Molecular Biology, Elsevier Science Publishing Co., New York, (1986)286.
6. Doyle, T., and D. Botstein, Proc. Natl. Acad. Sci. USA, 93 (1996) 3886.
7. Fischer M., S. Kaech, D. Knutti and A. Matus, Neuron, 20 (1998) 847.
8. Heidecker, M., Y. Yan-Marriott and Marriott, G., Biochemistry, 34 (1995) 11017.
9. Li J. C., Yen, Liaw D. K., Podsypanina, S. Bose, S.I. Wang, J. Puc, C. Miliaresis, L. Rodgers, R. McCombie, S.H. Bigner, BC. Giovanella, M. Ittmann, B. Tycko, H. Hibshoosh, MH. Wigler and R. Parsons., Science, 275 (1997) 1943.
10. Liao, J. C., L.J. Berg, and D.G. Jay, Photochem. Photobiol., 62 (1995) 923.

11. McKenna, N, J.B. Meigs and Y-L. Wang, J. Cell Biol., 100 (1983) 292.

12. Ormo, M., A.B. Cubitt, K. Kallio, L.A. Gross, R.Y. Tsien, and S.J. Remington, Science, 273 (1996) 1392.

13. Prassler, J., S. Stocker, G. Marriott, M. Heidecker, J. Kellermann, and G. Gerisch, Mol. Biol. Cell, 8 (1997) 83.

14. Rosenblatt, J., B.J. Agnew, H. Abe, J.R. Bamburg and T.J. Mitchison, J. Cell Biol., 136 (1997) 1323.

15. Sanders M.C. and YL. Wang, J. Cell Sci., 100 (1991) 771.

16. Sechi, A.S., J. Wehland, and J.V. Small, J. Cell Biol., 137 (1997) 155.

17. Symons M.H. and T.J. Mitchison, J. Cell Biol., 114 (1991) 503.

18. Taylor, D.L. and Y.L. Wang, Proc. Natl. Acad. Sci. USA, 75 (1978) 857.

19. Theriot, J.A. and T.J. Mitchison. Nature, 352 (1991)126.

20. Valles, A.M., B. Boyer, J. Badet, G.C. Tucker, D. Barritault. and J.P. Thiery, Proc. Natl Acad. Sci. (USA), 87 (1990) 1124.

21. Vigers, G.P., M. Coue, and J.R. McIntosh, J. Cell Biol., 107 (1988) 1011.

22. Westphal, M., A. Jungbluth, M. Heidecker, B. Mühlbauer, C. Heizer, J-M. Schwartz, G. Marriott, and G. Gerisch, Curr. Biol., 7 (1997) 176.

23. Choidas A. Jungbluth A. Sechi A. Murphy J. Ullrich A., and Marriott G., Eur. J. Cell Biol., 77 (1998) 81.

24. Rozelle AL. Machesky LM. Yamamoto M. Driessens MH. Insall RH., and Roth MG., Current Biology, 10 (2000) 311.

25. Defacque H. Egeberg M. Habermann A. Diakonova M. Roy C. Mangeat P. Voelter W. Marriott G. Pfannstiel J. Faulstich H., and Griffiths G., EMBO Journal, 19 (2000) 199.

Molecular Anatomy of Cellular Systems
I. Endo et al., (editors)
© 2002 Elsevier Science B.V. All rights reserved.

Golgi-acting drugs: inducers and inhibitors of Golgi dispersal as probes to analyze Golgi membrane dynamics

Akira Takatsuki

Animal and Cellular System Laboratory, RIKEN (The Institute of Physical and Chemical Research)
Hirosawa 2-1, Wako-shi, Saitama 351-0198, Japan

Pharmacological compounds have been proven useful for obtaining fundamental insights into anterograde and retrograde trafficking between the endoplasmic reticulum and the Golgi and also into Golgi membrane dynamics. This review summarizes recent work on inducers and inhibitors of Golgi dispersal. To make this review up-to-date, I intended to include our observations that are under review for publication. I reviewed the action of brefeldin A, nordihydroguaiaretic acid, and arachidonyltrifluoromethyl ketone as inducers of Golgi dispersal, and of phospholipase A_2 antagonists, PDMP, PPMP, and mepanipyrim as inhibitors of pharmacologically induced Golgi dispersal. Brefeldin A is an inhibitor of GDP/GTP exchange reaction on adenosine ribosylation factors. Both nordihydroguaiaretic acid and arachidonyltrifluoromethyl ketone are inhibitors of phospholipase A_2 and also induce Golgi dispersal. However, some other phospholipase A_2 antagonists inhibit Golgi dispersal induced by nordihydroguaiaretic acid and arachidonyltrifluoromethyl ketone. PDMP and PPMP are inhibitors of glucosylceramide synthase, but their action on pharmacologically induced Golgi dispersal seem to be dissected from their effect on sphingolipid metabolism. Therefore, the action on the Golgi seems to be independent of their reported target molecules and remain to be elucidated. The molecular target of mepanipyrim action has not been clarified.

1. INTRODUCTION

The Golgi apparatus has a striking architecture in mammalian cells. Allied to this extraordinary architecture, the Golgi apparatus plays a central role in the exocytic pathway [1]. Proteins newly synthesized in the endoplasmic reticulum (ER) are delivered to the cis Golgi network, and then transported, in sequence, through the cis, medial and trans cisternae of the Golgi apparatus to the trans Golgi network (TGN). In TGN, defined functionally as the exit face of the Golgi, proteins are sorted to their destinated places, such as the plasma membrane, secretory granules or endosomes. In addition to exocytic pathway from the ER to Golgi, cargo and vehicle molecules are retrogradely transported from the Golgi back to the ER. Transport proceeds by repeated cycles of formation and fuse of transport vesicles from donor and with acceptor membrane, respectively [2]. The molecular nature of these events have been unraveled using both biochemical and genetic approaches. Inhibitors such as N-ethylmaleimide and brefeldin A (BFA) have contributed in these studies [3- 5).

Many molecules are continually transported into and out of the Golgi, and thus the molecules that constitute this organelle are not in a static state. In addition, the Golgi breaks down at the onset of mitosis, undergoes extensive fragmentation, and merges with the ER [6, 7). As cells exit mitosis, small stacks reassemble and move back to rebuild the Golgi architecture seen in interphase cells. In marked contrast to the progress made in identifying molecules involved in traffic through the Golgi, only a very limited number of molecules have so far been identified that are involved in constructing this unique organelle, as pointed out by Rabouille and Warren [8). Disassemble and reassemble of the Golgi have been investigated using cells in mitosis and exiting mitosis, respectively [6-8). N-Ethylmaleimide [9) and microcystin [10) have been applied to these investigations. And, ilimaquinone has been used to disassemble the Golgi as a mimic of the disassembly of the Golgi in mitosis [6, 11).

Drugs are very limited that are useful in investigations of intracellular protein trafficking and Golgi membrane dynamics [5). When BHK cells are infected with Newcastle disease virus (NDV) and virus envelope glycoprotein that has activity to induce syncytium formation is expressed on the cell surface, giant syncytia are formed [12). Using this virus-cell system, we have selected drugs that block syncytium formation without affecting virus glycoprotein synthesis as candidates for protein trafficking inhibitors. The Golgi apparatus play crucial functions in protein trafficking, and therefore it seems to be reasonable to expect that Golgi-acting drugs should be concentrated among these inhibitors. In a hope to find Golgi-acting drugs, we have surveyed their action on the Golgi architecture, and disassembly and rebuilding of the Golgi. In this study, we have used the advantage of the reversible action of BFA on the Golgi. BFA is a macrolide antibiotic produced by various fungi. It was first reported to be an inhibitor of protein and nucleic acid synthesis [13), but we found its novel action, i.e. inhibition of protein trafficking without affecting its synthesis [14). BFA disintegrates the Golgi, and fragmented Golgi and TGN merge with the ER and endosome/lysosome, respectively [4, 15). This BFA action is readily reversible, and Golgi architecture is rebuilt after its removal. Protein trafficking inhibitors tested are grouped into those that disrupt Golgi architecture without effect on both merging with the ER and the rebuilding of the Golgi, and that affect the merge or rebuilding in addition to those not having any of these effects.

After the initiation of the present study, several pharmacological compounds were reported to induce retrograde Golgi-to-ER membrane flow and to inhibit BFA-induced Golgi membrane flow. Their action is compared in this review.

2. INDUCERS OF GOLGI DISPERSAL

Golgi membrane is fragmented and relocated to the ER and then merge with ER membrane on mitosis [6-8, 16). Cytosol prepared from cells in the mitotic phase of cell cycle induces Golgi membrane fragmentation. However, its mechanism still remains unclarified. Pharmacological compounds that mimic the fragmentation and merge of the Golgi in mitotic cells should be expected to be useful prove to shed light on the mechanism. Maintenance of the Golgi architecture depends on microtubules [17), and nocodazole and ilimaquinone, disruptants of microtubules, fragment the Golgi and disperse it throughout the cytoplasm [5].

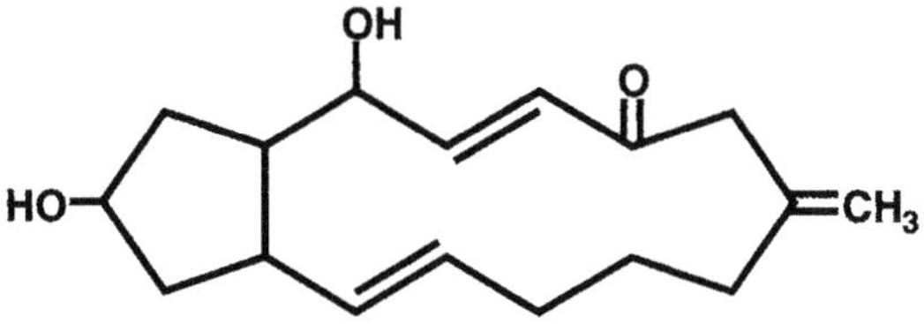

Figure 1. Structure of BFA

However, BFA, nordihydroguaiaretic acid (NDGA), and arachidonyltrifluoromethy ketone (AACOF$_3$) fragment the Golgi without affecting microtubules (see below). In addition to these three compounds, clofibrate and cyclofenil diphenol fragment the Golgi and induce retrograde Golgi-to-ER membrane flow [18, 19).

2.1. Brefeldin A (BFA)

BFA (Figure 1) is a macrolide antibiotic produced by diverse kinds of fungi. It was reported to inhibit protein and nucleic acid syntheses [13]. We found its novel action, i.e. blockade of intracellular protein trafficking without affecting its synthesis [14]. Later studies indicate that BFA inhibits GDP/GTP exchange reaction on adenosine ribosylation factors (ARFs), and as such prevents association of coat protein (β-COP) and blocks transport vesicle formation [20, 21]. Another prominent characteristic of BFA action is fragmentation of the Golgi and TGN and induction of their merge with the ER and lysosome/endosome, respectively [15]. The dissociation of coat protein from Golgi membrane is an early event in the induction of retrograde Golgi-to-ER membrane flow, and is thought to be a prerequisite for the induction of Golgi fragmentation [22]. However, dissociation of β-COP per se seems not to be the soul cause for Golgi fragmentation as shown by the observations that NDGA and AACOF$_3$ induce it without dissociating β-COP from Golgi membrane as described in the following. There are excellent reviews on BFA [4, 5), and readers are recommended to refer to them and references therein.

2.2. Nordihydroguaiaretic acid (NDGA)

NDGA (Figure 2) induces retrograde membrane flow to the ER from both the Golgi and TGN [19, 23). NDGA inhibits a diverse kind of enzymes such as lipoxygenase, phospholipase A$_2$, Na$^+$,K$^+$- and Mg^{2+}-adenosine triphosphatases, phosphofructokinase, luciferase, and α-glucosidase [24-27, and references in 19, 23). As predicted from this action, NDGA interacts with a diverse kind of proteins and its inhibitory action against α-glucosidase can be decreased heterogeneous proteins present in the reaction mixture [27]. Bovine serum albumin most effectively protected α-glucosidase from NDGA action. In agreement with this observation, presence of fetal calf serum in the medium interferes NDGA action to induce retrograde Golgi-to-ER membrane flow [27]. Depletion of fetal calf serum from the medium does not exert any detectable effect on cultured cells during a short period of incubation less than 2 hrs. Therefore, it is recommendable to use a medium not supplemented with serum in NDGA treatment.

Two prominent differences exist between the action of BFA and NDGA. One is that BFA induces a merge of the Golgi and TGN with the ER and lysosome/endosome, respectively [4, 15), while NDGA merges both with the ER [19). Another is that BFA dissociates β-COP

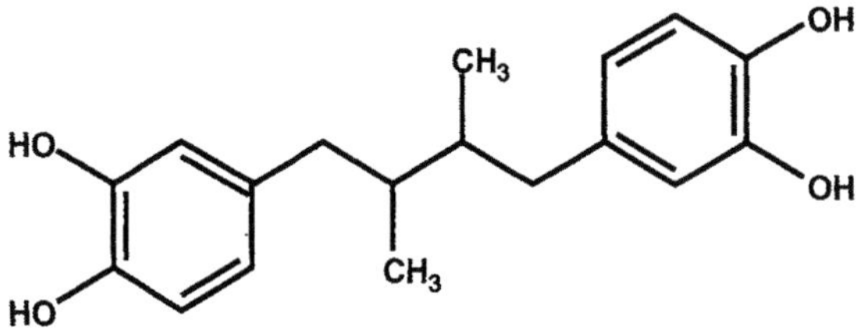

Figure 2. Structure of NDGA

from Golgi membrane [22), but NDGA does not [19, 23). The action of NDGA was correlated with its inhibition of lipoxygenase and/or phospholipase A_2. However, we now know that NDGA interacts with a diverse kind of proteins. Therefore, this correlation seems to require a reevaluation, taking its interaction with a diverse kind of proteins into consideration.

2.3. Arachidonyltrifluoromethy ketone (AACOF₃)

Phospholipase A_2 inhibitors disturb membrane trafficking and/or maintenance of the Golgi architecture [19, 23, 28-32). Stimulators of phospholipase A_2, melittin and phospholipase A_2-activating protein peptide, enhance cytosol-dependent Golgi membrane tubulation [32).

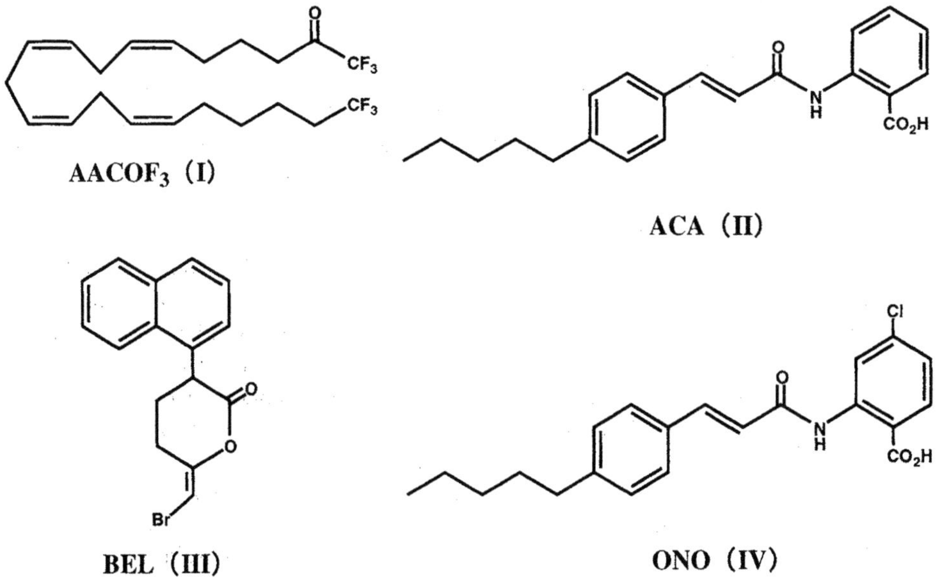

AACOF₃ (I)

ACA (II)

BEL (III)

ONO (IV)

Figure 3. Structure of phospholipase A_2 antagonists

Currently, our understanding of the mechanism(s) by which phospholipase A_2 regulates membrane trafficking and Golgi architecture is unclear. NDGA, a phospholipase A_2 inhibitor, induces fragmentation and vesiculation of both the Golgi cisternae and TGN, and disperse them throughout the cytoplasm as described above (cf. 2.2). AACOF$_3$ (Figure 3, I) is a phospholipase A_2 antagonist. It was initially reported to inhibit BFA-induced retrograde Golgi-to-ER membrane flow in clone 9 rat hepatocytes [28]. However, we found that AACOF$_3$ induces dispersal of the Golgi and TGN in a dose-dependent fashion [33]. Circumstantial evidences suggest that both the Golgi and TGN apparently seem to merge with the ER, but experimental supports for this wait for further studies.

Other phospholipase A_2 inhibitors, ACA (N-(p-amylcinnamoyl)anthranilic acid; Figure 3, II), BEL (E-6-(bromomethylene)-tetrahydro-3-(1-naphthalenyl)-2H-pyran-2-one; Figure 3, III), and ONO (2-(p-amylcinnamoyl)amino-4-chlorobenzoic acid; Figure 3, IV) also vesiculate the Golgi and TGN, but retain them in the juxtanuclear region (cf. 3.1) [28]. The latter phospholipase A_2 inhibitors retarded AACOF$_3$-induced dispersal of the Golgi cisternae and TGN. Induction of dispersal of the Golgi cisternae and TGN by phospholipase A_2 inhibitors, NDGA and AACOF$_3$, is not associated with dissociation of β-COP from membranes [19, 23, 33), while associated with BFA-induced dispersal [4, 22]. Despite of these similarities, the Golgi cisternae and TGN showed a different relative sensitivity to NDGA and AACOCF$_3$ (Table 1). The Golgi cisternae and TGN were dispersed at a similar time course of AACOF$_3$ treatment. However, the Golgi cisternae was dispersed at an earlier time of NDGA treatment (less than 3 min) than the TGN (about 10 min). TGN also showed less sensitivity (more than 30 min) to the action of BFA than ManII (less than 3 min). This characteristic of AACOF$_3$ clearly differentiates it from other well-known inducers of Golgi dispersion, BFA and NDGA. The organization of the Golgi cisternae and TGN depends on microtubule cytoskeleton [17]. AACOF$_3$ disintegrated the architecture of the Golgi cisternae and TGN at the same time in NRK cells as nocodazole, a microtubule-disrupting agent. However, differences were observed between their action. AACOF$_3$ did not disrupt microtubule. In addition, small vesicular ManII- and TGN38-stain was retained throughout the cytoplasm for more than 1 hr of nocodazole treatment, while they seemed to merge with the ER and/or endosome/lysosome after 10 min of AACOF$_3$ treatment as in the cases of BFA and NDGA treatment.

Table 1

Comparison of the action of BFA, NDGA, and AACOF$_3$ on the Golgi and TGN

	BFA	NDGA	AACOF$_3$
Relocation Site			
Golgi	ER	ER	not defined
TGN	Endosome / lysosome	ER	not defined
Relocation Time			
Golgi	< 3 min	< 3 min	~ 10 min
TGN	~ 30 min	~ 10 min	~ 10 min

3. INHIBITORS OF PHARMACOLOGICALLY INDUCED GOLGI DISPERSAL

As overviewed above, several pharmacological compounds were found to induce Golgi dispersal. However, very restricted are those which inhibit Golgi dispersal and/or retrograde Golgi-to-ER membrane flow. Recently, phospholipase A_2 antagonists and PDMP (D,L-threo-1-phenyl-2-decanoylamino-3-morpholino-1-propanol) were found to inhibit BFA-induced Golgi dispersal [28, 34]. We found PPMP (D,L-threo-1-phenyl-2-hexadecanoylamino-3-morpholino-1-propanol), an analogue of PDMP having a stronger activity against glucosylceramide synthase, and mepanipyrim disturb Golgi architecture and affect pharmacologically induced Golgi dispersal [35, 36]. Their action is briefly reviewed in the following.

3.1. Phospholipase A_2 antagonists
Phospholipase A_2 antagonists, such as AACOF$_3$, ACA, BEL, and ONO (Figure 3), inhibit BFA-induced Golgi-to-ER membrane flow in clone 9 rat hepatocytes [28]. However, their effect on BFA-induced Golgi dispersal is not significant in NRK cells [36]. We have at present no answer for this discrepancy, but cell-type specificity of their action might be one possible cause. However, they partly inhibit Golgi dispersal induced by NDGA, AACOF$_3$, or clofibrate [33, 36]. NDGA and AACOF$_3$ are inhibitors of phospholipase A_2. Therefore, these observations suggest that induction and blockade of Golgi dispersal by them are dissected from their inhibition of phospholipase A_2.

3.2. PDMP and PPMP
The action of PDMP (Figure 4) on BFA-induced retrograde Golgi-to-ER membrane flow was correlated with its action on calcium homeostasis [34]. However, PDMP blocks the BFA-induced ADP-ribosylation of BARS-50 in isolated Golgi membranes [37], and this is consistent with a role of ADP-ribosylation in the action of BFA and with the involvement of

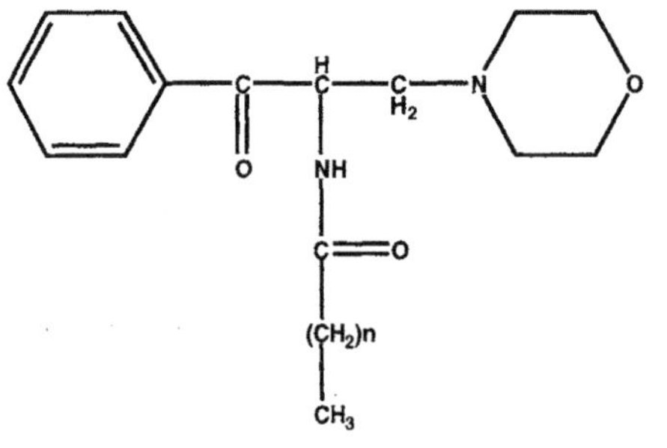

n=8 **PDMP** (2-Decanoylamino-3-morphilino-1-phenyl-1-propanol)
n=13 **PPMP** (2-Hexadecanoylamino-3-morphilino-1-phenyl-1-propanol)

Figure 4. Structure of PDMP and PPMP

BARS-50 in the regulation of membrane trafficking. In addition, inhibition of thymidine transport is a nonspecific effect of PDMP [38]. Therefore, the clarification of the mechanism of PDMP inhibition of BFA action on the Golgi apparatus remains open for further studies.

Kok et al. addressed the action of PDMP only to BFA-induced retrograde trafficking from the Golgi cisternae to ER [34]. Given the importance of the mechanism of maintenance of the Golgi architecture, we addressed whether or not BFA-, NDGA-, and nocodazole-induced Golgi disorganization was sensitive to PDMP, and also its effect on pharmacologically induced fragmentation of the TGN. In addition, to correlate sphingolipid metabolism with Golgi organization, action of PDMP was compared with that of PPMP, an analogue of PDMP having a stronger activity [39]. PDMP and PPMP blocked BFA- and NDGA-induced dispersal of the Golgi cisternae and TGN, and Golgi-derived vesicles were retained in the juxtanuclear region. PDMP and PPMP did not stabilize microtubules but blocked nocodazole-induced extensive fragmentation and dispersal of the Golgi, and large Golgi vesicles were retained in the juxtanuclear region. PPMP is a stronger inhibitor of glucosylceramide synthesis than PDMP, but they showed a reversed activity against BFA-induced retrograde membrane flow. However, PPMP showed a stronger activity as for Golgi disruption and inhibition of anterograde trafficking from the ER and rebuilding of the Golgi architecture. Cumulatively, these results suggest that sphingolipid metabolism is implicated in maintenance of the Golgi architecture and anterograde membrane flow from the ER but not in Golgi dispersal induced by BFA.

3.3. Mepanipyrim

Mepanipyrim (Figure 5), a pyrimidinamine, has been developed as a fungicide [40]. Mepanipyrim inhibits cell surface expression of VSV-G and C_6-NBD-glucosylceramide but not of C_6-NBD-sphingomyelin in BHK cells, and unesterified cholesterol accumulates in mepanipyrim-treated human skin fibroblasts [41]. However, until now there has been little experimental evidence as for its action mechanism. Pyrimenthanil [42] and CGA 219417 [43] are pyrimidinamine compounds structurally related to mepanipyrim. The principle of their action is assumed to be the inhibition of methionine biosynthesis [44, 45]. However, inhibition of virus multiplication by mepanipyrim could not be reversed in the presence of an excess amount of methionine (unpublished observation). We found that mepanipyrim inhibits BFA-induced retrograde protein trafficking from the Golgi to the ER. This inhibition is not restricted to BFA action, and mepanipyrim inhibits retrograde Golgi-to-ER trafficking induced by NDGA or clofibrate [36]. Mepanipyrim does not inhibit anterograde trafficking from the ER of Golgi-resident and newly synthesized proteins after the removal of BFA. Moreover, mepanipyrim distinguishes the Golgi cisternae from the TGN, and does not inhibit BFA- or NDGA-induced merge of the TGN with lysosome/endosome or the ER. Mepanipyrim does not stabilize microtubules, but prevent nocodazole-induced vesiculation and dispersion of the Golgi throughout the cytoplasm. These results suggest that the mepanipyrim-sensitive proteins participate in stabilizing the Golgi and its anchoring in the perinuclear region, and equally importantly, that mepanipyrim action to differentiate the TGN from the Golgi may be used as a pharmacological tool for investigating Golgi membrane dynamics.

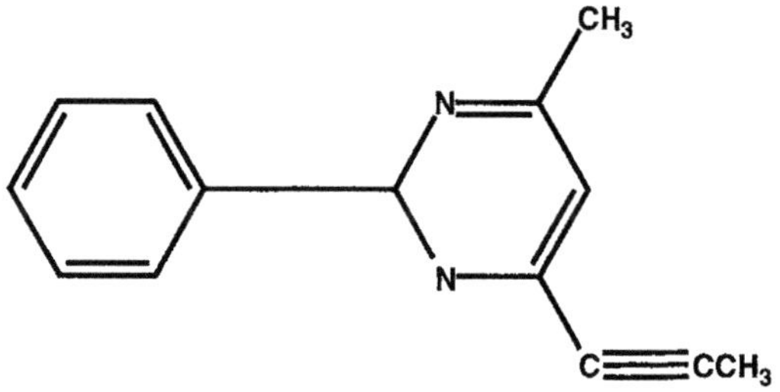

Figure 5. Structure of mepanipyrim

4. CONCLUDING REMARKS

Many proteins and lipids are transported into and out of the Golgi and TGN. In addition, the Golgi apparatus is disintegrated at mitotic phase of cell cycle and evenly partitioned into daughter cells, and then Golgi architecture is rebuilt in them. These observations indicate that the Golgi apparatus is an intracellular organelle in a dynamic equilibrium. However, the mechanism beneath this dynamic state of the Golgi apparatus has not been clarified as yet. Pharmacological blockade of Golgi dispersal or pharmacological induction of Golgi dispersal may be expected to shed light on the mechanism. Usefulness of pharmacological compounds in this field of investigation may be supported by the fact that about 200 papers, in which BFA was used as a probe, have been reported every year in the last decade. However, the mechanism action of the drugs introduced in this brief review including BFA has not been elucidated as yet. In addition, we have no evidence for that the induction of Golgi dispersal by pharmacological compounds mimics that in mitotic cells. Therefore, the next goal of investigations using these pharmacological compounds should aim at the elucidation of their action mechanism and also the clarification of the correlation, if any, between pharmacological and mitotic Golgi dispersal.

REFERENCES

1. M.G. Farquhar and G.E. Palade, Trends Cell Biol., 8 (1998) 2.
2. R. Jahn and T.C. Sudhof, Annu. Rev. Biochem., 68 (1999) 863.
3. J.E. Rothman and L. Orci, Nature, 355 (1992) 409.
4. R.D. Klausner, J.G. Donaldson and J. Lippincott-Schwartz, J. Cell Biol., 116 (1992) 1071.
5. A. Dinter and E.G. Berger, Histochem. Cell Biol., 109 (1998) 571.
6. F.A. Barr and G. Warren, Semin. Cell Dev. Biol., 7 (1996) 505.
7. J. Lippincott-Schwartz and K.J.M. Zaal, Histochem. Cell Biol., 114 (2000) 93.
8. C. Rabouille and G. Warren, in The Golgi Apparatus, (E. G. Berger and J. Roth eds.), Birkhauser Verlag, Basel, pp. 195, (1997).
9. C. Rabouille, T.P. Levine, J.M. Peters and G. Warren, Cell, 82 (1995) 905.
10. C. MacKintosh, K.A. Beattie, S. Klumpp, P. Cohen and G.A. Codd, FEBS Lett., 264, (1990) 187.

11. N. Cabrera-Poch, R. Pepperkok and D.T. Shima, Biochim. Biophys. Acta, 1404 (1998) 139.

12. M. Muroi, A. Takasu, M. Yamasaki and A. Takatsuki, Biochem. Biophys. Res. Commun. 193 (1993) 999.

13. V. Betina and L. Montagnier, Bull. Sco. Chim. Biol., 48 (1966) 194.

14. A. Takatsuki and G. Tamura, Agric. Biol. Chem., 49 (1985) 899.

15. N.W. Chege and S.R. Pfeffer, J. Cell Biol., 111 (1990) 893.

16. K.J.M. Zaal, C.L. Smith, R.S. Polishchuk, N. Altan, N.B. Cole, J. Ellenberg, K. Hirschberg, J.F. Presley, T.H. Roberts, E. Siggia, R.D. Phair and J. Lippincott-Schwartz, Cell, 99 (1999) 589.

17. J. Thyberg and S. Moskalewski, Exp. Cell Res., 159 (1985) 1.

18. P. de Figueiredo and W.J. Brown, Cell Biol. Toxicol., 15 (1999) 311.

19. D. Drecktrah, P. de Figueiredo, R.M. Mason and W.J. Brown, J. Cell Sci., 111 (1998) 951.

20. J .G. Donaldson, D. Finazzi and R.D. Klausner, Nature, 360 (1992) 350.

21. J.B. Helms and J.E. Rothman, Nature, 360 (1992) 352.

22. J. Scheel, R. Pepperkok, M. Lowe, G. Griffiths and T.E. Kreis, J. Cell Biol., 137 (1997) 319.

23. T. Fujiwara, N. Takami, Y. Misumi and Y. Ikehara, J. Biol. Chem., 273 (1998) 3068.

24. C.L. Armour, J.M. Hughes, J.P. Seale and D.M. Temple, Eur. J. Pharmacol., 72 (1981) 93.

25. M. Tagaya, N. Henomatsu, T. Yoshimori, A. Yamamoto, Y. Tashiro and T. Fukui, FEBS Lett., 324 (1993) 201.

26. G.L. Kellett, E.D. Barker, N.L. Beach and J.A. Dempster, Biochem. Pharmacol., 45 (1993) 1932.

27. K. Sato, M. Muroi, M. Nakamura, M. Fujimura and A. Takatsuki, Biosci. Biotechnol. Biochem., 65 (2001) 996.

28. P. de Figueiredo, D. Drecktrah, J.A. Katzenellenbogen, M. Strang and W.J. Brown, Proc. Natl. Acad. Sci. USA, 95 (1998) 8642.

29. M. Tagaya, N. Henomatsu, T. Yoshimori, A. Yamamoto, Y. Tashiro and S. Mizushima, J. Biochem. (Tokyo), 119 (1996) 863.

30. P. de Figueiredo, R.S. Polizotto, D. Drecktrah and W.J. Brown, Mol. Biol. Cell, 10 (1999) 1763.

31. T. Babia, I. Ayala, F. Valderrama, E. Mato, M. Bosch, J.F. Santaren, J. Renau-Piqueras, J.W. Kok, T.M. Thomsen and G. Egea, J. Cell Sci., 112 (1999) 477.

32. R.S. Polizotto, P. de Figueiredo and W.J. Brown, J. Cell. Biochem., 74 (1999) 670.

33. N. Kuroiwa, M. Nakamura, M. Tagaya and A. Takatsuki, Biochem. Biophys. Res. Commun., 281 (2001) 582.

34. J.W. Kok, T. Babia, C.M. Filipeanu, A. Nelemans, G. Egea and D. Hoekstra, J. Cell Biol., 142 (1998) 25.

35. M. Nakamura, N. Kuroiwa, Y. Kono and A. Takatsuki, Biosci. Biotechnol. Biochem., 65 (2001) 1369.

36. M. Nakamura, N. Kuroiwa, Y. Kono and A. Takatsuki, Biosci. Biotechnol. Biochem., 65 (2001) 1812.

37. M.A. De Matteis, A. Luna, G. DiTullio, D. Corda, J.W. Kok, A. Luini and G. Egea, FEBS Lett., 459 (1999) 310.

38. R.D. Griner and W.B. Bollag, J. Pharmacol. Exp. Ther., 294 (2000) 1219.

39. S. Radin, J.A. Shayman and J. Inokuchi, Adv. Lipid Res., 26 (1993) 183.

40. S. Maeno, I. Miura, K. Masuda and T. Nagata, Brighton Crop Prot. Conf. Pests Dis., 2 (1990) 415.

41. I. Miura, M. Muroi, N. Shiragami, I. Yamaguchi and A. Takatsuki, Biosci. Biotech. Biochem., 60 (1996) 1690.

42. G.L. Neumann, E.H. Winter and J.E. Pittis, Brighton Crop Prot. Conf. Pests Dis., 4 (1992) 395.

43. U.J. Heye, J. Speich, H. Siegle, A. Steinemann, B. Foster, G. Knauf-Beiter, J. Herzog and A. Hubele, Crop Prot., 13 (1994) 541.

44. P. Masner, P. Muster and J. Schmid, Pest. Sci., 42 (1994) 163.

45. R. Fritz, C. Lanen, V. Colas and P. Leroux, Pest. Sci., 49 (1997) 40.

Molecular Anatomy of Cellular Systems
I. Endo et al., (editors)
© 2002 Elsevier Science B.V. All rights reserved.

Regulation of protein sorting and trafficking between the endoplasmic reticulum and the Golgi apparatus in yeast

Akihiko Nakano

Molecular Membrane Biology Laboratory, RIKEN
Wako, Saitama 351-0198, Japan

Molecular mechanisms of protein transport and sorting during the vesicular traffic between the endoplasmic reticulum and the Golgi apparatus in yeast have been extensively studied. The roles of small GTPases and the retrieval receptor Rer1p will be discussed in detail. Some experimental attempts with a higher plant *Arabidopsis* will also be described as an approach to understanding the importance of vesicular traffic in multicellular organisms.

1. INTRODUCTION

For the architecture of eukaryotic cells, it is important to understand how membrane organelles are organized and how they interact with each other. They represent central problems in the Biodesign Research. Vesicular transport plays a pivotal role in these processes. In this chapter, I will describe several pieces of work that have been carried out in the Molecular Membrane Biology Laboratory, RIKEN. The budding yeast, *Saccharomyces cerevisiae*, has been used for most of the molecular mechanistic studies. We are also interested in how multicellular organisms depend on intracellular events for their architecture of bodies. Vesicular transport is important in the establishment and maintenance of cell polarity and thus contributes to tissue biogenesis. To address this problem, we have also undertaken some approaches with a higher plant, *Arabidopsis thaliana*.

2. MOLECULAR MECHANISMS OF VESICLE BUDDING FROM THE ER – WITH A FOCUS ON THE SMALL GTPase Sar1p

It is now well known that families of small GTPases play essential roles in the intracellular protein trafficking. For vesicular transport, two distinct families of GTPases exist: Sar/Arf and Rab/Ypt. Discoveries of these classes of GTPase were made in late 1980's through the studies of protein secretion in yeast *S. cerevisiae*. As the representative of Rab/Ypt proteins, Sec4p and Ypt1p were identified by the groups of Novick, Botstein and Gallwitz (1-3). Sar1p was isolated by myself (4). The findings that these small GTPases are essential for protein transport between secretory organelles stimulated the field of vesicular traffic very much and a huge numbers of studies followed. So far, hundreds of Sar/Arf and Rab/Ypt proteins have

46

been identified from not only yeast but also from animals and plants (5-8).

2.1. Sar1p, a key player in vesicle budding

SAR1, the gene encoding Sar1 GTPase, was first isolated as a multicopy suppressor of the *sec12* temperature-sensitive mutant. *sec12* is one of the conditional mutants isolated by Novick *et al.* (9) that show a temperature-sensitive defect in the transport from the endoplasmic reticulum (ER) to the Golgi apparatus in yeast. During the course of the cloning of *SEC12* by complementation, I found that a completely distinct gene can also complement *sec12* (10). The sequence of the second gene indicated that it encodes a Ras-like small GTPase, so I named it *SAR1* for a secretion-associated Ras-related gene (4). Further studies showed that *SAR1* is also essential for ER-to-Golgi transport and thus for growth. Now it is known that Sar1p is required for the assembly of a coat protein complex (COPII) during the vesicle budding process from the ER membrane (11).

2.2. Sar1p as a molecular switch

The paradigm is that GTPases are "molecular switches" (12-14). GTPases can take two different conformations according to the states of the bound nucleotide, GTP or GDP. The GTP-bound form is usually the "active" form and performs a certain biological job. The GDP-bound form is regarded as the "inactive" form in contrast. What is very important here is that the conversion between these two processes is not reversible. As shown in Fig. 1, nucleotide exchange from GDP to GTP is required for the activation, and the hydrolysis of GTP to GDP and Pi shuts down the activity of the protein. As the name indicates, GTPases have their intrinsic GTPase activities, but they are very low by themselves. Another type of proteins, called GTPase activating proteins (GAPs), are required for efficient hydrolysis of GTP. The other reaction, the exchange of GDP and GTP, also needs a catalyst. It is called a guanine-nucleotide exchange factor (GEF). Different GAPs and GEFs are required for different GTPases and thus

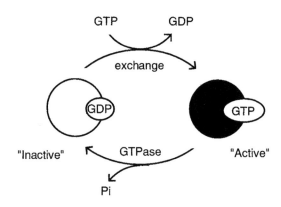

Fig. 1 GTPase cycle. The active and inactive states are interconverted by two different reactions.

comprise a very complex network of regulation. The targets of the active GTPases are known as effectors.

2.3. Regulation of the GTPase cycle of Sar1p

In the budding reaction from the ER, what is the role of Sar1 GTPase? What does the active Sar1p-GTP promote? There has been a big and long controversy on this problem. Because Sar1p-GTPγS or Sar1p-GMPPNP, complexes of Sar1p with a non-hydrolyzable GTP analog, can promote vesicle budding *in vitro* and *in vivo* (11, 15-17), it is certain that the turn-off of the Sar1 switch is not essential for budding itself. Then what is the physiological role of GTP hydrolysis for Sar1p? We have been proposing that the GTP hydrolysis is required for the uncoating reaction. That is, the conversion of Sar1p-GTP to Sar1p-GDP causes the release of Sar1p from the COPII vesicles, which leads to the complete disassembly of the COPII coat. In fact, Sar1p-GTPγS or Sar1p-GMPPNP is capable of forming vesicles but is completely defective in the overall transport to the Golgi and accumulates COPII-coated vesicles (11, 16).

Another possibility is that the hydrolysis of GTP is used as a way of proof reading for the selection of cargo molecules in the ER that are destined to the Golgi. If the sorting of cargo is not very good, Sar1p can hydrolyze GTP and abort budding. On the other hand, it is also plausible that the hydrolysis of GTP by Sar1p gives a commitment of vesicle formation so that the budding reaction reaches the point of no return.

Whichever model is correct, the GTP hydrolysis by Sar1p should not occur too early under the normal physiological conditions. It will give a timing of commitment. Here arises a paradox. It is now proven that for the case of Sar1p, Sec12p is the GEF (18) and Sec23p is the GAP (19) (Fig. 2). While the role of Sec12p as the GEF is reasonable as will be discussed later, the role of Sec23p GAP is puzzling. Sec23p is an essential component of the COPII coat. It is present as a part of the coat complex from the very beginning of vesicle budding. How come it suspends its GAP action onto Sar1p?

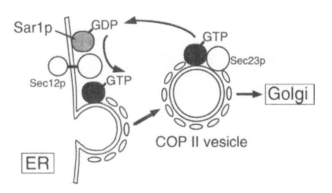

Fig. 2   The role of Sar1p GTPase cycle in the budding of a COPII vesicle from the ER membrane.

2.4.Sed4p as a novel regulator of Sar1p

We reasoned that some other component(s) in the ER would inhibit the GAP activity of Sec23p until the formation of COPII vesicles is completed. In support of this, Matsuoka et al. (20) reported that Sar1p-GTP can drive vesicle budding from the ER membrane in vitro but not the budding from liposomes. Some component(s), most probably in the ER membrane, may be inhibiting the GAP activity of Sec23p. We became interested in the role of Sed4p in this context. Sed4p is a paralog of Sec12p, the GEF of Sar1p, but has no GEF activity towards Sar1p. Genetic evidence indicates that Sed4p acts as an activator of Sar1p. It is the best suppressor of sar1 ts mutants (21). Furthermore, it aggravates the ts defect of sec23 when overexpressed (22). We examined the effect of purified Sed4p on the GAP activity of Sec23p and found that it in fact showed some inhibition. The extent of inhibition of not good enough to explain the suspension of GTP hydrolysis in vivo but gives a good hint that a complex containing Sed4p may have a role in vivo in the regulation of GTP hydrolysis of Sar1p (22).

48

2.5.Arf1p, a cousin of Sar1p, performs a multiple job

The Sar/Arf family of GTPases consists of Sar1p and many Arf proteins. Whereas the role of Sar1p is very well conserved from yeast to higher eukaryotes in the COPII vesicle budding from the ER, the understanding of Arf functions has been quite controversial. We noticed that this is because most of the studies on Arfs were done in vitro but not in vivo, and decided to undertake an in vivo genetic approach with yeast. We screened for temperature-sensitive alleles of yeast ARF1 (23) in the _ arf2 background and identified 8 new ts mutants. Their detailed characterization revealed that the single Arf1 protein of yeast performs a surprisingly various functions (24). They show defects in ER-to-Golgi anterograde transport, Golgi-to-ER retrograde transport, Golgi-to-vacuole and endosome-to-vacuole transport and so on in an allele-dependent fashion. Furthermore, these mutants could complement each other in some combinations of heterozygotes. These results indicate that Arf1p is indeed a multifunction protein unlike Sar1p (24). Functional differentiation between different Arfs will be a next interesting problem to be addressed.

3. SORTING OF MEMBRANE PROTEINS IN THE PROTEIN TRAFFIC BETWEEN THE ER AND THE GOLGI APPARATUS

Not only the cargo but also resident proteins in the ER and Golgi must be sorted from one another during the transport processes. Sorting of membrane proteins is especially important from the viewpoint of correct functioning of transport machinery. The dilysine motif has been known as the ER localization signal for a subset of membrane proteins, but there are many that do not harbor this signal. Mechanistic understanding of membrane protein sorting remains mostly in the black box.

3.1. Static retention versus dynamic retrieval – lessons from Sec12p

Since the very early days of the analysis of Sec12p, we have been very much interested in its behavior. Since it is the GEF for Sar1p (18), it must reside in the ER and in fact it does, but the majority of the molecules are subject to N- and O-linked oligosaccharide modifications in the Golgi (10). This suggested its recycling between the ER

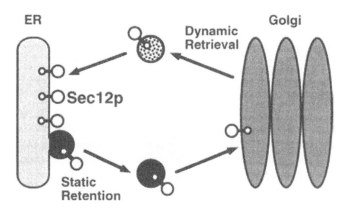

Fig. 3   Two mechanisms of ER localization of Sec12p. Static retention prevents exit of Sec12p from the ER and dynamic retrieval sends back Sec12p from the Golgi to the ER.

and the Golgi (10). However, we also realized that Sec12p is hardly detected in the purified COPII vesicles forming *in vitro* (16). There must be another mechanism that prevents Sec12p from entering the vesicles. We postulated that Sec12p is localized to the ER by two mechanisms, static retention in the ER and dynamic retrieval from the Golgi (Fig. 3).

3.2 Isolation of *rer* mutants – identification of molecular components

To prove this hypothesis and identify molecular machinery involved in these two mechanisms, we screened for mutants that are defective in the correct ER localization of Sec12p (25). We isolated two different types of mutants and named them *rer1* and *rer2* (standing for retention in the ER or retrieval to the ER). Cloned wild-type *RER1* and *RER2* genes code for an integral membrane protein in the Golgi and a peripheral membrane protein in the ER, respectively (26, 27).

3.3. Two signals in Sec12p

While the localization of Rer1p in the Golgi suggested its role in the retrieval, we further asked whether Sec12p has two different signals for the two mechanisms. An extensive study using chimeric molecules showed that Sec12p in fact harbors two signals: one for static retention in the N-terminal cytoplasmic domain and the other for dynamic retrieval in the transmembrane domain (Fig. 4) (28). The latter signal works completely in an Rer1p-dependent manner, supporting the role of Rer1p in retrieval (28).

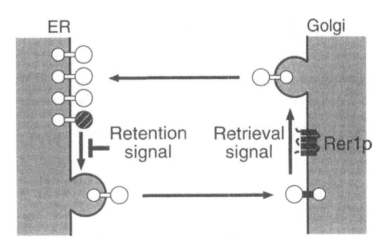

Fig. 4 Sec12p has two ER localization signals: one in the cytoplasmic domain for retention and the other in the transmembrane domain for Rer1p-dependent retreical.

3.4 Rer1p as general sorting machinery involved in ER-Golgi recycling

To examine whether the role of Rer1p is specific for Sec12p or not, we looked at other ER resident membrane proteins. To our surprise, Sec71p and Sec63p, which take completely different topology from Sec12p, also depend on Rer1p for the ER localization (29). We further showed that Sec12p and Sec71p compete for the Rer1p-dependent retrieval, which suggested that a saturable mechanism operates for this retrieval (29).

3.5. Rer1p is a retrieval receptor

The direct physical interaction between Sec12p and Rer1p was examined by crosslinking experiments. They are crosslinked only when the *bona fide* retrieval signal exists in the TMD of Sec12p (30). This is the first proof that Rer1p acts as a sorting receptor for the ER retrieval signal. We further showed that the C-terminal tail of Rer1p is essential for the Golgi localization. The COPI coat recognizes signal(s) in this tail and controls its correct localization probably through very dynamic vesicle recycling not only between the ER and the Golgi but also within the Golgi and perhaps between the Golgi and endosomes (30).

3.6. Rer2p – a legend of dolichol

As Rer2p is in the ER, we first suspected that it may be involved in the retention mechanism. Further studies, however, unveiled a big surprise. Rer2p turns out to be the *cis*-prenyltransferase, which is a long-searched key enzyme of the dolichol biosynthesis (27). This finding was made in parallel with the observation that its *E. coli* homolog, *rth*, is essential for the synthesis of cell wall peptidoglycan through the formation of a lipid carrier (31). As the missorting defects of *rer2* cannot be explained simply by the deficiency of glycosylation, we suggest that dolichol may have a novel physiological role in the membrane of the ER. Yeast has another *cis*-prenyltransferase, Srt1p. We found that the synthesis of Rer2p and Srt1p is differently regulated over the yeast life cycle (32). Srt1p is highly expressed only in the stationary phase. Furthermore, many enzymatic characteristics are different. Srt1p makes longer polyprenols than Rer2p and resides in lipid particles (32).

3.7. Regulation of Sec12p function – the most upstream event

Since the nucleotide exchange of Sar1p by Sec12p is the earliest event in the vesicle budding from the ER as far as we know, it is very interesting to know how the Sec12p function could be regulated. It may link the sorting events in the lumen of the ER with budding. To address this question, we have screened for novel genes that genetically interact with the *sec12* ts mutant (33). Among the extragenic suppressor mutations, we identified *HRR25*, a yeast gene encoding a casein kinase I (34). The loss of function of *HRR25* suppresses the transport defect of *sec12*, suggesting that the Hrr25 kinase is negatively regulating the Sec12p function. Sec12p itself is not a substrate of Hrr25p. To search for molecules that link Sec12p and Hrr25p, we are now performing a two-hybrid screening which gives several interesting candidates. Analysis of multicopy suppressors of *sec12* is also going on.

4. APPROACHES TO HIGHER PLANTS

While the yeast *S. cerevisiae* is an ideal model organism to learn molecular mechanisms of vesicular transport, it is also interesting to consider the role of membrane traffic in multicellular organisms. In the studies of mammalian cell lines which establish polarity in culture, evidence is accumulating that vesicular traffic is very important for the development and maintenance of cell polarity. It is also needless to say that higher-order functions of mammals, such as neuronal signal transmission and antigen presentation, depend very much on vesicular traffic processes. In contrast, little is known about membrane trafficking in higher plants. Because plant cells cannot move in tissues in general, the morphogenesis of plant bodies should be virtually determined by the directions of cell division and cell elongation. These two processes require vesicular processes. We have recently started projects to address these problems using higher plants *Arabidopsis* and tobacco.

4.1.Sar1 GTPase of *Arabidopsis thaliana* and *Nicotiana tabacum*

A very powerful approach to understanding the roles of particular genes in plant physiology is genetics. Both forward and reverse genetics are feasible with higher plants, especially with *Arabidopsis thaliana*. However, because the secretory functions are essential for growth in yeast, plant mutants that are defective in vesicular transport may well be lethal. To get around this problem, the best way is to try to isolate conditional mutants. Dominant mutations are especially useful because they can confer phenotypes even in diploid cells. So, we decided to establish a system, in which a dominant mutation of a certain secretory gene can be turned on conditionally.

From our studies with yeast, we know that *SAR1* is one of the most useful tools to manipulate vesicular transport. We have identified many dominant mutations in yeast (17, 35, 36). So, we first tried to isolate plant counterparts of yeast *SAR1* and identified *AtSAR1* from *A. thaliana* and *NtSAR1* from tobacco *Nicotiana tabacum* (37). These *SAR1* homologs were able to complement the lethality of yeast Δ*sar1* cells. Dominant mutations were designed from the knowledge of yeast *SAR1* and tested for their effects on the wild-type genes. Several mutations were shown to confer dominant negative effects to either yeast *SAR1* or corresponding plants wild-type genes for the cell growth and ER-to-Golgi transport in yeast (37).

The next step was to express these mutant plant *SAR1* genes in plant cells in a conditional fashion. After a long struggle, we were finally able to establish a transient expression system with *Arabidopsis* suspension culture cells and the tobacco BY-2 cell line. As markers of transport, we have made fusion proteins with green fluorescent protein (GFP), such as AtRer1B-GFP (38), AtErd2-GFP and Sporamin-GFP. Upon the transient expression of *AtSAR1* H74L, both Golgi markers (AtRer1B-GFP and AtErd2p-GFP) and the vacuole marker (Sporamin-GFP) changed their localization to the ER (39). This is the first proof that plant *SAR1* genes function in the ER-to-Golgi transport. Furthermore, this approach provides us with wonderful possibilities of the manipulation of membrane trafficking in plants. Similar experiments with Arabidopsis *ARF* genes are now underway.

4.2.Other small GTPases and their functional analysis

In contrast to the Sar/Arf family of GTPases, which are required for vesicle budding, Rab/Ypt GTPases are involved in vesicle targeting and fusion processes. Among many plant members of this GTPase family, we have been working on the *Arabidopsis ARA* genes (8).

In the process of testing whether any of *ARA* genes can complement yeast *ypt* mutants, we found that some of them were rather toxic to yeast mutant cells. *ARA4* gave clearest results. When the wild-type or Q71L mutant of *ARA4* was expressed, *ypt1*, *ypt3* and *sec4* mutant cells were not able to grow any more even at the permissive temperature. We reasoned that this was probably due to the titration of common regulators of Rab/Ypt proteins in yeast. To test this possibility, we screened an *Arabidopsis* cDNA library for clones that can suppress the deleterious effect of *ARA4* Q71L in the yeast *ypt1* mutant. The strongest suppressor clone encoded a plant counterpart of the Rab GDP dissociation inhibitor (GDI), which we named *AtGDI1* (40). *AtGDI2* was also identified by homology (41). GDI has been known as a common regulator of Rab/Ypt proteins, which detaches Rab-GDP from the membrane and keeps it in the cytosolic pool. The same suppressor screen led to isolation of another *Arabidopsis* novel gene, which we named *SAY1* (42). The precise role of *SAY1* is not clear at the moment, but using the yeast *ypt* mutant cells as a kind of test tube, we could examine the modes of interaction between the Ara4 protein and its regulators in detail. A very similar approach was undertaken for the Pra2 GTPase from pea, which also caused a growth defect in *ypt* mutants, but in this case GDI was not the component being titrated in yeast cells (43).

Very recently we have identified a completely new type of Rab protein from *Arabidopsis*, Ara6. It differs from conventional Rab/Ypt proteins in many respects. Most remarkably, it lacks the C-terminal Cys motif which is usually geranylgeranylated in other Rab proteins, but harbors myristoylation and palmitoylation sites at the N-terminus. This type of Rab GTPases have been found only in higher plants, suggesting that plants have developed a trafficking system completely different from yeast and animal cells. The detailed analysis of Ara6 is now underway.

4.3. Other approaches in plants

In a long run, forward genetics to isolate mutants that are defective in membrane traffic will prove fruitful. However, to set up a screening, we have many more problems to solve. For example, markers of transport are still very limiting. We have made a variety of GFP-fusion proteins and are now testing their usability for screening. In one of such processes, we found that the vacuolar membranes of *Arabidopsis* show very complex and dynamic structures. Its physiological meanings are now being investigated.

To identify trafficking machinery in plants, we can take advantage of the knowledge in yeast in a variety of ways. Complementation of yeast *sec* mutants is a very straightforward way and we have identified an *Arabidopsis* gene, *RMA1*, which complements the yeast *sec15* mutant (44). Its product contains a RING finger motif and a C-terminal membrane anchor domain. As yeast does not have a true counterpart of this gene, the function of Rma1 in plants remained a big puzzle to us. But very recently, we could show that it is a novel type of ubiquitin ligase (45). Its physiological role in *Arabidopsis*, especially in membrane trafficking, would be the next question to be solved.

5. ACKNOWLEDGMENT

The work described in this chapter was mainly carried out by the former and present members of Molecular Membrane Biology Laboratory of RIKEN. They include: Hiroshi Abe, Ken Sato, Akiko Murakami, Miyuki Sato, Yumiko Saito-Nakano, Masaki Takeuchi, Noriyuki Matsuda, Takashi Ueda, and Chieko Saito. Their enthusiasm is highly appreciated.

REFERENCES

1. A. Salminen and P. Novick, J. Cell Biol., 109 (1987) 1023.
2. N. Segev, J. Mullholand and D. Botstein, Cell, 52 (1988) 915.
3. H. D. Schmitt, M. Puzicha and D. Gallwitz, Cell, 53 (1988) 635.
4. A. Nakano and M. Muramatsu, J. Cell Biol., 109 (1989) 2677.
5. S. Ferro-Novick and P. Novick, Annu. Rev. Cell Biol., 9 (1993) 575.
6. C. Nuoffer and W. E. Balch, Annu. Rev. Biochem., 63 (1994) 949.
7. D. P. S. Verma, C.-I. Cheon and Z. Hong, Plant Physiol., 106 (1994) 1.

8. H. Uchimiya, T. Anai, E. T. Aspuria, M. Matsui, A. Nakano and T. Ueda, J. Plant Res., 111 (1998) 257.
9. P. Novick, C. Field and R. Schekman, Cell, 21 (1980) 205.
10. A. Nakano, D. Brada and R. Schekman, J. Cell Biol., 107 (1988) 851.
11. C. Barlowe, L. Orci, T. Yeung, M. Hosobuchi, S. Hamamoto, N. Salama, M. F. Rexach, M. Ravazzola, M. Amherdt and R. Schekman, Cell, 77 (1994) 895.
12. Y. Kaziro, Biochim. Biophys. Acta, 505 (1978) 95.
13. H. R. Bourne, D. A. Sanders and F. McCormick, Nature, 348 (1990) 125.
14. Y. Kaziro, H. Itoh, T. Kozasa, M. Nakafuku and T. Satoh, Annu. Rev. Biochem., 60 (1991) 349.
15. T. Oka, S. Nishikawa and A. Nakano, J. Cell Biol., 114 (1991) 671.
16. T. Oka and A. Nakano, J. Cell Biol., 124 (1994) 425.
17. Y. Saito, K. Kimura, T. Oka and A. Nakano, J. Biochem., 124 (1998) 816.
18. C. Barlowe and R. Schekman, Nature, 365 (1993) 347.
19. T. Yoshihisa, C. Barlowe and R. Schekman, Science, 259 (1993) 466.
20. K. Matsuoka, L. Orci, M. Amherdt, S. Y. Bednarek, S. Hamamoto, R. Schekman and T. Yeung, Cell, 93 (1998) 263.
21. Y. Saito, T. Yamanushi, T. Oka and A. Nakano, J. Biochem., 125 (1999) 130.
22. Y. Saito-Nakano and A. Nakano, Genes Cells, 5 (2000) 1039.
23. T. Stearns, R. A. Kahn, D. Botstein and M. A. Hoyt, Mol. Cell. Biol., 10 (1990) 6690.
24. N. Yahara, T. Ueda, K. Sato and A. Nakano, Mol. Biol. Cell, 12 (2001) 221.
25. S. Nishikawa and A. Nakano, Proc. Natl. Acad. Sci. USA, 90 (1993) 8179.
26. K. Sato, S. Nishikawa and A. Nakano, Mol. Biol. Cell, 6 (1995) 1459.
27. M. Sato, K. Sato, S. Nishikawa, A. Hirata, J. Kato and A. Nakano, Mol. Cell. Biol., 19 (1999) 471.
28. M. Sato, K. Sato and A. Nakano, J. Cell Biol., 134 (1996) 279.
29. K. Sato, M. Sato and A. Nakano, Proc. Natl. Acad. Sci. USA, 94 (1997) 9693.
30. K. Sato, M. Sato and A. Nakano, J. Cell Biol., 152 (2001) 935.
31. J. Kato, S. Fujisaki, K. Nakajima, Y. Nishimura, M. Sato and A. Nakano, J. Bacteriol., 181 (1999) 2733.
32. M. Sato, S. Fujisaki, K. Sato, Y. Nishimura and A. Nakano, Genes Cells, 6 (2001) 495.
33. A. Nakano, J. Biochem., 120 (1996) 642.
34. A. Murakami, K. Kimura and A. Nakano, J. Biol. Chem., 274 (1999) 3804.
35. A. Nakano, H. Ohtsuka, M. Yamagishi, E. Yamamoto, K. Kimura, S. Nishikawa and T. Oka, J. Biochem., 116 (1994) 243.
36. T. Yamanushi, A. Hirata, T. Oka and A. Nakano, J. Biochem., 120 (1996) 452.
37. M. Takeuchi, M. Tada, C. Saito, H. Yashiroda and A. Nakano, Plant Cell Physiol., 39 (1998) 590.
38. K. Sato, T. Ueda and A. Nakano, Plant Mol. Biol., 41 (1999) 815.
39. M. Takeuchi, T. Ueda, K. Sato, H. Abe, T. Nagata and A. Nakano, Plant J., 23 (2000) 517.
40. T. Ueda, N. Matsuda, T. Anai, H. Tsukaya, H. Uchimiya and A. Nakano, Plant Cell, 8 (1996) 2079.
41. T. Ueda, T. Yoshizumi, T. Anai, M. Matsui, H. Uchimiya and A. Nakano, Gene, 206 (1998) 137.

42. T. Ueda, N. Matsuda, H. Uchimiya and A. Nakano, Plant J., 21 (2000) 341.
43. N. Matsuda, T. Ueda, Y. Sasaki, and A. Nakano, Cell Struct. Funct., 25 (2000) 11.
44. N. Matsuda and A. Nakano, Plant Cell Physiol., 39 (1998) 545.
45. N. Matsuda, T. Suzuki, K. Tanaka and A. Nakano, J. Cell Sci., 114 (2001) 1949.

Molecular Anatomy of Cellular Systems
I. Endo et al., (editors)
© 2002 Elsevier Science B.V. All rights reserved.

An unexpected gift from fungicide metabolism studies: blasticidin S deaminase (BSD) from *Aspergillus terreus*

Makoto Kimura[a], Masaki Yamamoto[b], Makio Furuichi[b], Takashi Kumasaka[b], and Isamu Yamaguchi[a]

[a] Microbial Toxicology Laboratory, RIKEN
2-1 Hirosawa, Wako, Saitama 351-0198, Japan
[b] Structural Biophysics Laboratory, RIKEN
1-1-1 Kouto, Mikaduki, Sayo, Hyogo 679-5143, Japan

The *bsd* gene is now being used worldwide as a drug resistance marker gene in the field of molecular cell research. This useful tool is an unexpected gift from the fungicide metabolism studies of 30 years ago. At the time blasticidin S deaminase (abbreviated as BSD) was purified from *Aspergillus terreus*, it was never expected that we are still dealing with it. Here we briefly summarize the "history" of studies on BSD and "preview" some interesting topics in our current research.

1. DISCOVERY OF BSD FROM *ASPERGILLUS TERREUS*

blasticidin S deaminase (BSD)
EC 3.5.4.23

blasticidin S (BS)   deaminohydroxy-blasticidin S (d-BS)

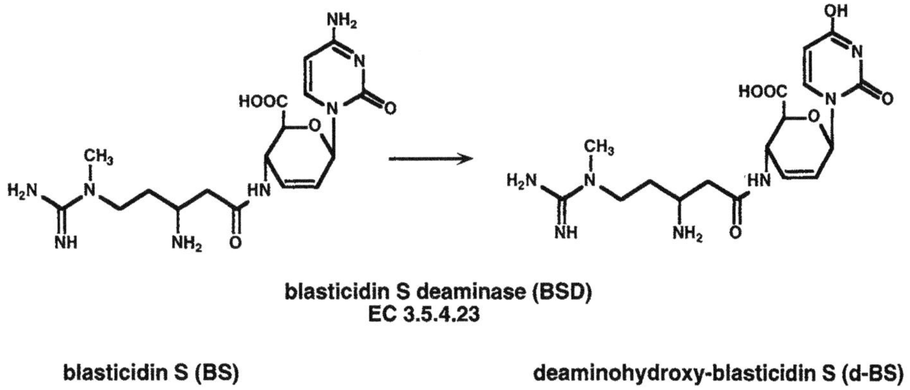

Figure 1. Inactivation of BS by BSD from *A. terreus*.

Blasticidin S (BS) is the first successful fungicide of microbial origin developed in Japan [1]. BS is a potent protein synthesis inhibitor of both prokaryotes and eukaryotes, and belongs to a group of aminoacylnucleoside antibiotics. It has widely been used as a replacement for the phenylmercurial fungicide to control rice blast disease caused by *Pyricularia oryzae* (perfect stage; *Magnaporthe grisea*).

The metabolic fate of BS in the environment was investigated in our laboratory in the late 1960's. It was found that this antibiotic is first converted to biologically inactive deaminohydroxy-BS (Fig. 1), and then further degraded to unknown substances.

A microorganism responsible for the deamination reaction was isolated from a paddy of soil and identified as a fungal strain of *Aspergillus terreus* S-712. An enzyme responsible for this reaction, named BSD, was purified from the fungus and its enzymatic properties were examined [2]. BSD exhibited high specificity toward BS and some of its analogues. No deamination occurred toward other purine nucleosides, such as cytosine, cytidine, and deoxycytidine.

2. ISOLATION OF THE BS RESISTANT GENE (*BSR*) FROM A BACTERIAL STRAIN OF *BACILLUS CEREUS*

Endo and co-workers have fortuitously found a bacterial strain that is able to grow in the medium containing > 500 µg/ml of the antibiotic. They identified the bacterium as *Bacillus cereus* strain K55-S1 and demonstrated that the resistance is coded by a genetic element on plasmid pBSR8 [3]. The strain K55-S1 also possessed the BS-deaminase activity as was the case in *A. terreus*.

Because of the broad inhibition spectrum of BS to many organisms, the resistance gene was expected to be useful as a selective marker gene in genetic manipulation of both prokaryotes and eukaryotes. For this purpose, the BS resistance gene, *bsr*, was isolated from pBSR8 [4] and used to transform various organisms. The *bsr* gene was useful in genetic transformation of *Escherichia coli*, *Nicotiana tabacum* [5], and HeLa cells [6], but it did not work in the fungus *P. oryzae*. BS was especially excellent in inhibiting the growth of HeLa cells, which saved time and efforts required in the selection process [6]. Based on this result, BS has gradually attracted attentions of molecular cell biologists as an alternative reagent of G418.

3. CLONING OF *BSD* FROM *ASPERGILLUS TERREUS* - A SET OF DIFFERENT MARKER GENES WITH THE SAME FUNCTION

The cDNA encoding BSD was obtained by expression cloning in *E. coli*, and its nucleotide sequence was determined [7]. In comparison of *bsd* and *bsr* sequences, there were no *overall* similarities at both nucleotide and peptide sequence levels. However, a local region of limited similarity *was* found at the peptide sequence level between these two BS-deaminases (*i.e.*, the region that covers residues 85-95 and 97-107 of BSD and BSR, respectively). This region was also highly conserved among other members of the cytosine nucleoside/nucleotide deaminase family (Fig. 2). Recent crystallographic studies of *E. coli* cytidine deaminase (CDA) have shown that the region belongs to a part of a novel zinc-binding motif, by which a catalytic zinc is accommodated at the active center [8].

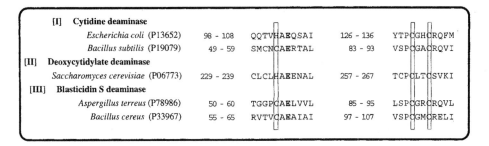

Figure 2.   Alignments of zinc-binding amino acid residues (boxed) in members of the cytosine nucleoside/nucleotide deaminase family.   The active sites Glu is shown in bold.

In addition to the absence of sequence similarity between *bsd* and *bsr*, a large difference in preference of codon use was observed for some amino acids.   In consistent with the fact that *Bacillus* is a typical low G + C content bacterium, *bsr* appeared to be optimized for expression in *B. cereus* and had a relatively high A + U content; *i. e.*, 74.3 % of codons end either A or U at the third position..   An extremely biased codon use was observed for Ile, Tyr, Gln, and Cys, in which > 93 % of the codons end in A or U.   In contrast, *bsd* does not have such a preference in codon use.   This result was important for efficient expression of the deaminase activity in various organisms [9].

As expected, *bsd* was useful as a marker gene in *P. oryzae* that could not be transformed to BS resistance by *bsr* [7].   Moreover, *bsd* proved to work much better than *bsr* for use as a marker gene in transfection of FM3A cells [10].   Based on these results, the *bsd*-BS selection system is now commercialized by Invitrogen Co. (Carlsbad, CA) for use in several organisms, including yeasts, mammalian cells, and insect cells.

4.   BIOCHEMICAL CHARACTERIZATION OF BSD

We established an efficient overproduction-purification system for BSD using the cDNA cloned from *A. terreus*.   The estimated M.W. of the purified enzyme indicated that BSD is a tetramer.   On SDS-PAGE, BSD was electrophoresed much slower if heat treatment of samples was not done before electrophoresis; *i.e.*, it was detected as a single band of 36 kDa [11].   When BSD was heated in SDS and then electrophoresed, it migrated according to its subunit size of 13 kDa.   Since the CD spectrum and BSD activity were substantially unaffected at similar concentrations of SDS, this heat-modifiable behavior is most likely to represent insufficient denaturation of the protein by SDS.

As predicted from the presence of the catalytic zinc-binding motif, BSD proved to contain one zinc per subunit by inductively coupled plasma (ICP) optical emission spectroscopy. However, the catalytic function appeared not to be the only role of this zinc for the enzyme. First, titration of the zinc-chelating -SH groups with p-hydroxymercuriphenylsulfonate led to dissociation of a BSD tetramer.   Second, depletion of zinc in reconstitution of chemically denatured BSD resulted in improper folding of the polypeptide.   Third, mutagenesis of Cys-91, one of the proposed zinc-binding residues of BSD, resulted in gross perturbation in enzyme structure. These results are suggestive of structural role of zinc in maintenance of the protein structure.

In alcohol dehydrogenases (ADHs), there are both types of zinc atom, one catalytic and one structural, per subunit of the enzyme. A catalytic zinc is coordinated to *three* amino acid residues in the protein and an activated water molecule, whereas a structural zinc is coordinated to *four* Cys residues in the conserved -C-X_2-C-X_2-C-X_7-C- sequence [12]. The zinc of BSD coordinated to *three* Cys residues thus appears to be quite unique in that it has both catalytic and structural functions.

5. STRUCTURE OF BSD

BSD was crystallized from precipitant solutions containing PEG 8000 and magnesium ions in the absence of BS [13]. The crystal structure was determined at 1.8 Å by the multi-wavelength anomalous diffraction (MAD) technique at RIKEN beamline I of Spring-8 (Yamamoto *et al.*; manuscript in preparation). The tetramer was related by 222 molecular symmetry and each subunit had identical folding pattern (Fig. 3).

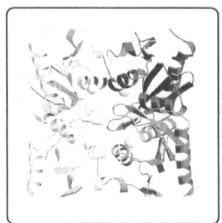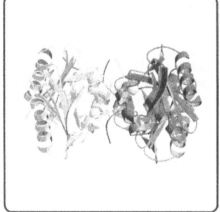

Figure 3. Two nearly orthogonal views of the BSD tetramer. The top view (left) and side view (right). The zinc atoms are shown as gray spheres.

The overall architecture of a subunit comprised of a six-stranded β-sheet sandwiched either side by α-helices, which was quite similar to that of the catalytic domain of *E. coli* CDA. The exact positions of the last four C-terminal residues could not be assigned due to the very weak electron densities even at the cryo-conditions. This short flexible region is predicted to cover the entrance channel of the active site and appears to play critical roles for substrate recognition.

In *E. coli* CDA, the His-102 residue was shown to be involved in zinc binding by X-ray crystal structure analysis. The corresponding residue is substituted by Cys-54 in BSD (Fig. 2), and as expected, zinc proved to form a tetrahedral coordination with the three Cys residues (*i.e.*, Cys-54, Cys-88, and Cys-91) and a water molecule as a nucleophile. Preliminary results indicate that the coordination sphere of the zinc atom is apparently similar to that reported for *E. coli* CDA. Further analysis on the behavior of zinc upon substrate binding is now under study to learn more details of the reaction mechanism of the zinc enzyme with the three-Cys coordination motif (Fig. 2).

6. FUTURE INTERESTS

Although the peptide sequence of BSD and CDA are rather divergent, structure analysis revealed that these enzymes share similar overall architecture and active center. Accepting that common origin can be inferred from protein structure conservation even their peptide sequences were quite divergent [14, 15], these enzymes may have descended from a common ancestral origin. In Southern analysis, however, there were no apparent *bsd* homologs in the genome of *A. terreus* (our unpublished observation). Thus the scenario for the evolution of the enzyme could involve multiple events of duplication, gene transfer, and/or selective loss. Phylogenetic analyses of BSDs (if any) and CDAs from *Aspergilli* and other fungal genera, as well as approaches based on comparative genomics, might provide hints to answer this question.

Obviously the reaction catalyzed by BSD is not a physiologically important process in biological systems. However, BSD *does* have several attractive characters and provide interesting topics as a model enzyme for pure basic protein research; *i.e.*, availability of good stable crystals for high resolution structure analysis, behavior of zinc during the enzyme reaction, and roles of the short flexible C-terminal region in catalysis. Out of such scientific curiosities, we are *still* "playing" with this enzyme, and the "history" of studies on BSD continues.

REFERENCES

1. K. Fukunaga, T. Misato, I. Ishii, and M. Asakawa, Bull. Agric. Chem. Soc. Jpn., 19 (1955) 181.
2. I. Yamaguchi, H. Shibata, H. Seto, and T. Misato, J. Antibiot., 28 (1975) 7.
3. T. Endo, K. Kobayashi, N. Nakayama, T. Tanaka, T. Kamakura, and I. Yamaguchi, J. Antibiot., 41 (1988) 271.
4. T. Kamakura, K. Kobayashi, T. Tanaka, I. Yamaguchi, and T. Endo, Agric. Biol. Chem., 51 (1987) 3165.
5. T. Kamakura, K. Yoneyama, and I. Yamaguchi, Mol. Gen. Genet., 223 (1990) 332.
6. M. Izumi, H. Miyazawa, T. Kamakura, I. Yamaguchi, T. Endo, and F. Hanaoka, Exp. Cell Res., 197 (1991) 229.
7. M. Kimura, T. Kamakura, Q.Z. Tao, I. Kaneko, and I. Yamaguchi, Mol. Gen. Genet., 242 (1994) 121.
8. L. Betts, S. Xiang, S.A. Short, R. Wolfenden, and C.W. Carter Jr., J. Mol. Biol., 235 (1994) 635.
9. M. Kimura and I. Yamaguchi, Pestic. Biochem. Physiol., 56 (1996) 243.

10. M. Kimura, A. Takatsuki, and I. Yamaguchi, Biochim. Biophys. Acta, 1219 (1994) 653.
11. M. Kimura, S. Sekido, Y. Isogai, and I. Yamaguchi, J. Biochem., 127 (2000) 955.
12. B.L. Vallee and D.S. Auld, Biochemistry, 29 (1990) 5647.
13. M. Nakasako, M. Kimura, and I. Yamaguchi, Acta Crystallogr. D Biol. Crystallogr., 55 (1999) 547.
14. A.G. Murzin, Curr. Opin. Struct. Biol., 6 (1996) 386.
15. L. Holm and C. Sander, Science, 273 (1996) 595.

Molecular Anatomy of Cellular Systems
I. Endo et al., (editors)
© 2002 Elsevier Science B.V. All rights reserved.

Isolation and analysis of genes from phytopathogenic fungi

Takayuki Motoyama[a], Tsutomu Arie[ab], Takashi Kamakura[a], and Isamu Yamaguchi[a]

[a] Microbial Toxicology Laboratory, RIKEN
2-1 Hirosawa, Wako, Saitama 351-0198, Japan
[b] Faculty of Agriculture, Tokyo University of Agriculture and Technology
3-5-8 Saiwaicho, Fuchu, Tokyo 183-8509, Japan

Control of phytopathogenic fungi is important for plant protection, particularly in humid climate like in Japan, where plant diseases cause serious damage to crop production. Many phytopathogenic fungi make infection-specific organs. Factors involved in differentiation and functioning of the organs are target of fungicides. We isolated many genes expressed during early stage in infection-specific organ formation by the cDNA subtraction technique (section 1). Inhibition mechanism of a fungicide, which interfere with the function of an infection-specific organ, was analyzed by the X-ray crystallographic study of a complex between the fungicide and the target enzyme. Structure-based analysis of the target enzyme showed involvement of a C-terminal region in substrate-binding process (section 2).

Mating event among phytopathogenic fungi is important because this process is related to evolution of pathohogenicity. We report isolation and analysis of the mating type genes, which determine the mating ability, from phytopathogenic fungi (section 3).

1. GENES EXPRESSED DURING EARLY STAGE OF APPRESSORIUM FORMATION BY THE RICE BLAST FUNGUS

The conidial germ tube of the rice blast fungus, *Pyricularia oryzae* (teleomorph: *Magnaporthe grisea*), differentiates an infection-specific structure, an appressorium, for penetration into the host plant. Formation of the appressorium is also observed on synthetic solid substrata such as polycarbonate. We found that a plant lectin, concanavalin A, specifically suppressed the appressorium formation without affecting the germling adhesion if it was applied within 2-3 hours after germination. Standing on the result, we constructed a cDNA library that represents the early stage of germ tube development and/or appressorium formation from the 2.5-hour-old germ tubes using a cDNA subtraction strategy by the combination of the biotin labeled driver method and adapter-primed PCR method. Out of 686 colonies of the library, 158 distinct clones' nucleotide sequences were partially analyzed. Some clones' expression patterns were detected by RT-PCR and from those results, our library seemed to well represent the objective developmental stage of *P. oryzae*.

1.1. Timing dependent inhibitory effects of concanavalin A

We have previously reported that a plant lectin, concanavalin A (ConA), inhibits appressorium formation without preventing the adhesion of germ tubes onto the solid surface [1], when ConA was applied after the conidial attachment to the solid surface occurred as described by Hamer et al.[2]. To estimate the timing of RNA isolation for construction of differential library, timing dependent inhibitory effects of ConA were observed. Germination was observed on most of the conidia within 1 h of incubation, and small swellings were found at tips of approximately 70% of the germlings at 2 to 4 h after the start of incubation. Inhibitory effect of ConA on appressorium formation was observed when the lectin was added to the germling no later than 2.5 h after incubation began. This phase-dependent effect was coincident with the microscopical observation of the starting period of swelling. Thus we estimated that 2 to 2.5 h after the start of incubation could be the crucial time for *P. oryzae* starting differentiation of appressorium.

1.2. Enrichment of cDNAs specifically expressed during the early stage of appressorium formation by cDNA subtraction

cDNA (0.2 µg) prepared from the germ tubes under appressorium forming condition (*AF*) was subtracted once or twice with more than 37 times excess of cDNA (7.5 µg) prepared from vegetative growth hyphae (*VG*) (Fig. 1). Single subtraction caused small differences in the PCR resultant, though the sample subtracted twice showed significant reduction of smear products and produced a distinguishable band pattern. The primer set for this PCR experiment was confirmed not to amplify the genomic DNA of *P. oryzae* and *VG*-cDNAs that did not have synthetic adapters. Thus we infer that this large reduction of PCR products reflects the efficient enrichment of *AF*-specific cDNA by the subtraction. PCR products from the double subtraction were cloned and then 686 colonies were stored as a grid differential library. The redundancy of the library was estimated by colony hybridization or comparison of 3' end sequences when we accumulated the end sequence data as described below. 250 clones' colony hybridization and 3' end sequence data showed that the redundancy of the library was roughly one third which suggested the number of distinct families in our grid library to be about 200, and 158 distinct clones have been identified already.

1.3. Detection of the differential expression by RT-PCR

For RT-PCR, both ends' DNA sequence of candidates A2 (plate A, colony number 2), A3, A4, A27, and A31 were analyzed and specific primers (30mer each) were synthesized. Total RNA from *AF* and *VG* conditions were prepared and 1 µg each of total RNA was reverse-transcribed into 1st strand cDNA and used for the PCR. To distinguish the genes the expression of which is affected by appressorium differentiation from genes generally expressed during the early stage of conidial germination, another condition (*NA*: non-appressorium-forming condition on a solid surface) was set for this RT-PCR analysis. Total RNA of an *NA* sample was prepared like the *AF* sample except that the droplet of conidia was prepared with 2% Yeast Extract solution, which was reported to inhibit appressorium induction [3]. Since the *VG* sample was prepared from a liquid culture, some genes specifically expressed during germination on the solid surface might not be involved in the *VG* sample and those genes could be important because gene products related to the sensing and/or intracellular signaling must be preset before switching on the appressorium

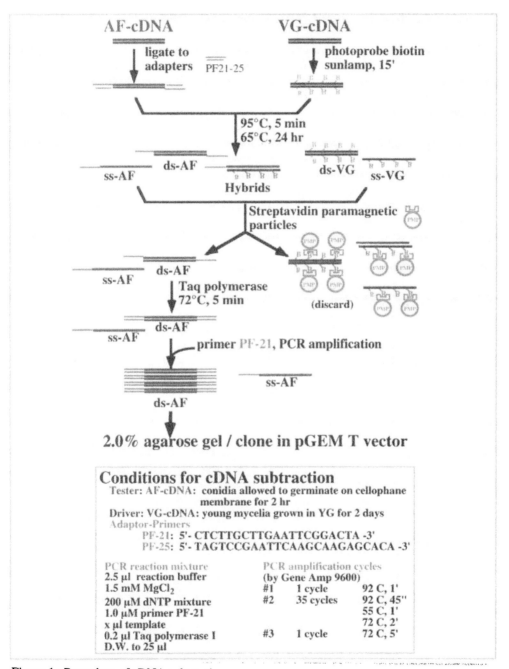

AF-cDNA **VG-cDNA**

ligate to adapters PF21-25

photoprobe biotin sunlamp, 15'

95°C, 5 min
65°C, 24 hr

ss-AF ds-AF ds-VG ss-VG

Hybrids

Streptavidin paramagnetic particles

ss-AF ds-AF

Taq polymerase
72°C, 5 min

(discard)

ss-AF ds-AF

primer PF-21, PCR amplification

ss-AF

ds-AF

2.0% agarose gel / clone in pGEM T vector

Conditions for cDNA subtraction
Tester: AF-cDNA: conidia allowed to germinate on cellophane
membrane for 2 hr
Driver: VG-cDNA: young mycelia grown in YG for 2 days
Adaptor-Primers
PF-21: 5'- CTCTTGCTTGAATTCGGACTA -3'
PF-25: 5'- TAGTCCGAATTCAAGCAAGAGCACA -3'

PCR reaction mixture
2.5 µl reaction buffer
1.5 mM MgCl₂
200 µM dNTP mixture
1.0 µM primer PF-21
x µl template
0.2 µl Taq polymerase I
D.W. to 25 µl

PCR amplification cycles
(by Gene Amp 9600)
#1	1 cycle	92 C, 1'
#2	35 cycles	92 C, 45"
		55 C, 1'
		72 C, 2'
#3	1 cycle	72 C, 5'

Figure 1. Procedure of cDNA subtraction.

differentiation. Clones A4 and A31 gave differential expression patterns. Clone A3 did not show a differential pattern. Clone A2 seemed to have slight differences between *AF* and *VG* but it was ambiguous and not conclusive by this RT-PCR analysis, though every RT-PCR experiment was repeated at least three times and patterns were reproducible for every clone tested. The sequence data of the 5' end of A3 had significant similarity to cyclophilinA, which is a molecular chaperon and known as a typical constitutive gene [4]. Templates prepared for this RT-PCR analysis were estimated using *MPG1* primer as a control and *MPG1* showed a similar pattern with Northern hybridization data which could be detectable with our RNA samples because it is highly abundant in this developmental stage [5]. According to those differential patterns given with primers for cyclophilin (clone A3) and *MPG1*, this RT-PCR system could detect the differential expression property to some extent. Clone A4 seemed to be completely repressed in the *VG* condition but expressed equally under the *NA* condition, so that clone A4 could be a specific gene which is expressed during early germination stage but not affected by appressorium induction. Clone A31 seems to have differences among *AF*, *VG*, and *NA* conditions, so that A31 may be a downstream gene of initiation of appressorium formation. Four more candidates were analyzed by RT-PCR and three showed differential expression patterns (data not shown). Since more than half of the candidate clones to be challenged by RT-PCR showed differential expression, it is strongly suggested that our subtraction strategy was successful in condensing the stage-specific cDNAs from the germ tubes in the early stage of development.

1.4. Partial end sequence analysis of the differential library

5' and 3' end sequences of part of the library clones were read with one-pass reading. 3' data were used only for detecting the overlapping clones and 5' data were used to search for similar sequences. The 5' end sequence of clone A4, which was expressed during germination, did not have similarity with any known genes. A31, which was expressed differentially between appressorium inducing and non-inducing conditions, showed moderate similarity in deduced amino acid sequence with *gEgh16*, which was cloned from germinating conidia of *Erysiphe graminis* by subtractive hybridization [6]. However, we found that a different clone, B15, was a real homolog of *gEgh16* and there were many differences between A31 and *gEgh16* as follows. A31 showed clear differential expression detected by RT-PCR. *gEgh16* was expressed less specifically. Although A31 was a single copy gene, *gEgh16* belongs to a multigene family. No similarity was found in DNA sequences of A31 and *gEgh16*, though B15 had significant DNA sequence similarity with *gEgh16*. Thus, A31 could have structural properties similar to *gEgh16*, which is probably a secreted glycoprotein, but the function might be different.

So far, 158 distinct clones' ends were sequenced. Among other clones which were not analyzed by RT-PCR, there were some apparent constitutive genes such as ribosomal proteins or histone H2B. Like cyclophilin (clone A3), those were considered to be 'escape' clones from subtraction. There were some signal transduction related gene homologs such as phospholipaseC, protein kinase, and GTP binding protein. Recently, signal transduction pathways related to the appressorium differentiation have been one of the most developing research fields on *P. oryzae* (reviewed by Dean [7]). None of our homologs of those genes were the same genes that had already been reported [8-12]. We also found two known genes of *P. oryzae*. One was a polyhydroxynaphthalene reductase gene, which is required for melanin biosynthesis to complete the formation of functional appressoria [13]. The other was a *Pwl1* homolog and *Pwl1* is known as an *Avr* gene [14]. Since both genes' products must exist or work during the infection, it is reasonable to be expressed in the germ tube, and at the same time, the existence of those genes could guarantee the quality of our differential library.

Detecting mRNA expressed during the early stage of germ tube development was extremely difficult. However, this stage must be very important to understand the early plant-microbe interaction which involves the induction of appressorium formation. Most of the cloned genes of *P. oryzae* related to the signal transduction were isolated based on the similarity with known genes of other organisms [9-11,15]. Gene products related to the initiation of appressorium differentiation are probably preset in or on the germ tube and work upstream of the known signal transduction pathways. The library constructed in this study is the earliest stage representing library to date and will be extremely useful to search for the genes working upstream of the intracellular signal transduction pathways.

2. FUNCTIONAL ANALYSIS OF SCYTALONE DEHYDRATASE FROM THE RICE BLAST FUNGUS

Scytalone dehydratase (STD), a melanin biosynthetic enzyme, is indispensable for the pathogenicity of the rice blast fungus, *P. oryzae*. Carpropamid ((1*RS*, 3*SR*)-2,2-dichloro-*N*-[1-(4-chlorophenyl)ethyl]-1-ethyl-3-methylcyclopropanecarboxamide) was developed as a potent controlling agent against the rice blast fungus. Enzyme kinetics data showed that carpropamid is a tight-binding competitive inhibitor of STD. From the X-ray crystal structure analysis of STD complexed with carpropamid, we identified interactions that determine the tight-binding. Structural data also inferred that the C-terminal region (154G-172K) is

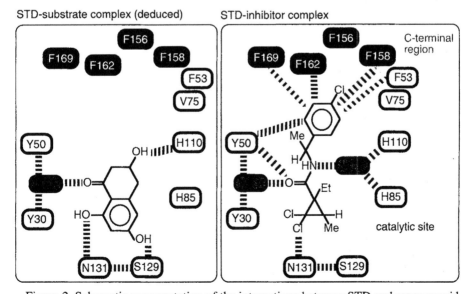

Figure 2. Schematic representation of the interactions between STD and carpropamid.

important for the enzyme function. We constructed STD-mutants at the C-terminal region as well as at other amino acid residues by the site-directed mutagenesis. Enzyme kinetics data for the STD-mutants with or without carpropamid suggested involvement of the C-terminal region in enzyme function and in the inhibitor binding.

2.1. Tight-binding inhibition of STD by carpropamid

cDNA of STD was cloned and overexpressed in *E. coli*. The purified recombinant STD had specific activity similar to that produced in *P. oryzae* [16] and used in further study. Of the four isomers of carpropamid, only KTU3616B (1*S*, 3*R*, 1'*R*) is active in melanin biosynthesis inhibition in intact blast cells and in blast control [17]. So, we analyzed STD-inhibition by using KTU3616B. Inhibition of STD by KTU3616B was observed at very low concentrations of the inhibitor close to the enzyme concentration, suggesting that KTU3616B is a tight-binding inhibitor of STD. We determined the inhibition type and the dissociation constant (K_i) for KTU3616B by fitting the enzymatic activity data to the specific equations developed for tight-binding inhibitors [18]. Predominant type of inhibition of STD by KTU3616B was competitive. The calculated K_i value was 140 pM, which is more than 105 times smaller than the K_m for scytalone.

2.2.3D-structure of a STD-carpropamid complex

To know the interactions that determine such a tight-binding inhibition, we then analyzed the structure of a complex of carpropamid and STD by an X-ray crystallographic study [19].

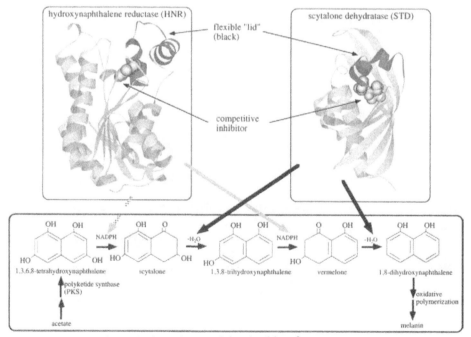

Figure 3. Melanin biosynthetic pathway of the rice blast fungus.

Crystals of the complexes were observed around pH 5.2 when PEG3350 was used as a precipitant. A structural model of the complex was obtained at 2.1 Å resolution by a cryogenic X-ray crystallographic analysis with use of a crystal. In particular, three types of interactions were considered to be important (Fig. 2). First, a strong hydrogen bond was observed between Asn 131 and chlorine. Second, two hydrogen bonds mediated by two water molecules were found. Third, the phenyl group of carpropamid interacted with aromatic amino acid residues of STD.

X-ray crystallographic study showed that carpropamid was completely embedded in STD (Fig. 3). This means that structural change of STD is needed for inhibitor binding and, probably, for substrate binding. From structural information, the part of structural change seems to be a C-terminal region (154G-172K, Fig. 3. black portion) because only the C-terminal region is flexible enough to enable access of molecules to the active site pocket. The C-terminal region may act as a flexible lid which control access of substrates and inhibitors to the active site pocket [19]. Interestingly, similar flexible region was found in a hydroxynaphthalene reductase (Fig. 3) [20,21], another enzyme in the melanin biosynthetic pathway. We then analyzed the function of the flexible C-terminal region of STD.

2.3. Structure-based analysis of STD

Deletion analysis of the C-terminal region showed that enzymatic activity was absent when at least 15 amino acids were deleted [16]. This result suggested that the C-terminal portion of STD was important for catalysis or structural integrity (or both). C-terminal region and some amino acid residues involved in the inhibitor binding were exchanged by site-directed mutagenesis and its effect on enzyme activity and inhibitor binding were analyzed (Fig. 4).

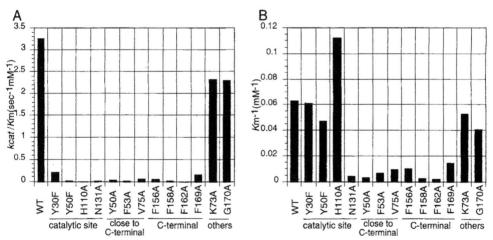

Figure 4. Functions of STD C-terminal region for enzyme reaction (A) and for substrate-binding (B).

Enzyme kinetics data for the STD-mutants indicated that all the amino acid residues involved in carpropamid-binding (Y30F, Y50F, H85A, H110A, N131A, Y50A, F53A, V75A, F158A, F162A, and F169A) were also important for enzyme function, although amino acid residues not involved in carpropamid binding (K73A and G170A) were not. Interestingly, all the STD with mutations at the C-terminal region (F156A, F158A, F162A, and F169A) or the region close to the C-terminal region (Y50A, F53A, and V75A) showed higher Km values than WT STD. However, mutations at the catalytic site (Y30F, Y50F, and H110A) afforded no such phenomenon. These data suggest the importance of the C-terminal region in substrate-binding process.

The Y30F, V75A, and F169A mutants retained higher enzyme activity , more than 5% of STD activity compared to the wild type enzyme, than other STD-mutants (less than 1% of activity). The Ki values for carpropamid increased in V75A and F169A, suggesting the specific involvement of V75 and F169 in carpropamid binding. However, the binding between STD and carpropamid was still tight, and no point mutation which may produce carpropamid-resistant strain was found.

Although reductase inhibitors of fungal melanin biosynthesis, such as tricyclazole, pyroquilon, and fthalide, have been used as anti-blast chemicals, no dehydratase inhibitors except carpropamid have been used for disease control yet. Detailed analyses of the function of the C-terminal region of STD may help both understanding the inhibition mechanism of STD-inhibitors and designing more efficient ones.

3. MATING TYPE GENES FROM SEXUAL AND ASEXUAL ASCOMYCETE FUNGI

We have established a polymerase chain reaction (PCR)- based strategy to clone mating type genes from ascomycetes. The approach was efficient for cloning mating type (*MAT*) loci from the heterothallic (sexual) pyrenomycete, *Gibberella fujikuroi*. Using the same strategy, we have also demonstrated that two asexual phytopathogenic ascomycetes, *Fusarium oxysporum* (a pyrenomycete) and *Alternaria alternata* (a loculoascomycete) also carry a MAT locus which has only a few structural differences compared to that of their heterothallic relatives. *MAT* genes from *F. oxysporum* and *A. alternata*, were expressed and *A. alternata* *MAT* genes were functional in the close heterothallic relative, *Cochliobolus heterostrophus*, when assayed by heterologous expression. We also propose a PCR-based method which is useful for determining the mating type of field isolates.

3.1. What are mating type genes?

Mating (sexual reproduction) is thought to be an important way for fungi to maintain genotypic variation that may allow adaptation to environmental changes [22]. In considering phytopathogenic ascomycetes, the development of a new pathogenic race compatible with a disease-resistant cultivar or the development of a new fungicide resistant strain may be due to genetic recombination during mating. No perfect stage has been documented in many phytopathogenic species such as *Fusarium oxysporum* and *Alternaria alternata*. Our objective is to determine the mechanisms of asexuality in asexual ascomycetes from a molecular viewpoint. As a first step in asking the question "What causes asexuality?", we used a PCR-based strategy to clone and compare mating type genes from heterothallic and asexual ascomycetes.

A single, regulatory locus *MAT* controls mating in heterothallic ascomycetes [23,24]. Sexual reproduction is initiated when strains of opposite mating type interact. *MAT* loci have been cloned from many heterothallic ascomycete fungi such as *Neurospora crassa* [25,26], *Podospora anserina* [27], and *Cochliobolus heterostrophus* [28]. The MAT locus has alternate sequences in each mating type termed idiomorphs that lack significant sequence similarity to each other [24,29]. All ascomycete *MAT* idiomorphs encode proteins with DNA-binding motifs, suggesting that *MAT* genes encode transcriptional regulators and control the expression of genes required for sexual reproduction [24]. In this report, we have defined the strains carrying the idiomorph encoding a protein with an alpha-box motif as MAT1-1 and the idiomorph encoding high mobility group (HMG-box) motif as MAT1-2 (Table 1).

3.2. Polymerase chain reaction (PCR) -based strategy for cloning mating type genes

The conserved HMG-box domain found in the MAT1-2-1 protein was used as a starting point for cloning and sequencing the entire *MAT1-2* idiomorph plus flanking regions. Degenerate primers designed to the HMG-box domain of *N. crassa* and *C. heterostrophus* were useful in amplifying this region from pyrenomycetes and loculoascomycetes, respectively [30]. To obtain the complete *MAT1-2* sequence, TAIL-PCR and/or inverse-PCR approaches were used with the primers designed to both flanking regions to amplify the opposite *MAT1-1* idiomorph [30,31]. The method was successful in cloning mating type genes from *Setosphaeria turcica* [30], *A. alternata* [31], *Gibberella fujikuroi* [32,33], and *F. oxysporum* [31].

3.3. Mating type genes from *Gibberella fujikuroi*

The *MAT* locus was cloned from heterothallic *G. fujikuroi* [32,33]. The *MAT1-1* idiomorph (4605 bp) carries 3 genes (*MAT1-1-1*, *MAT1-1-2*, and *MAT1-1-3*) and the *MAT1-2* idiomorph (3824 bp) carries a gene (*MAT1-2-1*) [33]. The *MAT1-1-1* gene encodes a protein (382 aa) with an alpha-box domain, *MAT1-1-2* encodes a protein (433 aa) with an alpha helix domain, and *MAT1-1-3* encods a protein (174 aa) with an HMG-box domain; *MAT1-2-1* gene encodes a protein (223 aa) with an HMG-box domain (Table 1) [33].

Table 1

New designition of mating type and mating type genes based on the structure of
MAT locus in ascomycete fungi

		Mating type and mating type gene				
		MAT1-1			MAT1-2	Reference
		MAT1-1-1 [a]	*MAT1-1-2* [b]	*MAT1-1-3* [c]	*MAT1-2-1* [d]	
Loculoascomycetes						
Cochliobolus heterostrophus	Heterothallic	MAT1-1			MAT1-2	28
		MAT1-1	no	no	*MAT1-2*	
Alternaria alternata	Asexual	MAT1-1			MAT1-2	31[h]
		MAT1-1-1	no	no	*MAT1-2-1*	
Pyrenomycetes						
Neurospora crassa	Heterothallic	matA			mata	25,26
		matA-1	*matA-2*	*matA-3*	*mata-1*	
Podospora anserina	Heterothallic	mat-			mat+	27
		FMR1	*SMR1*	*SMR2*	*FPR1*	
Magnaporthe grisea	Heterothallic	MAT1-2			MAT1-1	36
		?[e]	?	?	?	
Gibberella fujikuroi [f]	Heterothallic	-			+	32[h],33[h]
		MAT-1-1	*MAT-1-2*	*MAT-1-3*	*MAT-2*	
Fusarium oxysporum	Asexual	MAT1-1			MAT1-2	31[h]
		MAT1-1-1	*MAT1-1-2*	*MAT1-1-3*	*MAT1-2-1*	
Discomycetes						
Pyrenopeziza brassicae	Heterothallic	MAT1-2			MAT1-1	37
		PAD1	*PMT1* [g]	*PHB2*	*PHB1*	

[a] Encodes a protein with an alpha-box motif.

[b] Pyrenomycetes and Discomycetes only; Encodes a protein with an amphipathic alpha- helix motif.

[c] Pyrenomycetes and Discomycetes only; encodes a protein with an HMG-box motif, not homologous to *MAT1-2-1* HMG-box.

[d] Encode a protein with an HMG-box motif.

[e] No sequence and gene structure data has been published.

[f] Mating population A; Anamorph, *Fusarium moniliforme*.

[g] No alpha-helix motif but codes metallothioneinlike protein [37].

[h] Obtained in this research.

3.4.Mating type genes from asexual species

MAT loci were cloned from *F. oxysporum*, an asexual relative of *G. fujikuroi*, and from *A. alternata*, an asexual relative of *C. heterostrophus* [31]. The *MAT1-1* and *MAT1-2* idiomorphs were 4618 and 3850 bp in *F. oxysporum* and 1906 and 2220 bp in *A. alternata*, respectively. The structure of the idiomorphs is very similar to those of their heterothallic relatives, *G. fujikuroi* and *C. heterostrophus* (Table 1). Reverse transcriptase(RT) -PCR showed all *MAT* genes from both species were expressed [31,33].

A. alternata MAT genes were confirmed to be functional in a close sexual relative, *C. heterostrophus*, by heterologous expression (Figure 5, 6) [31]. This demonstrates that asexuality in *A. alternata* is not due to non-functional MAT genes.

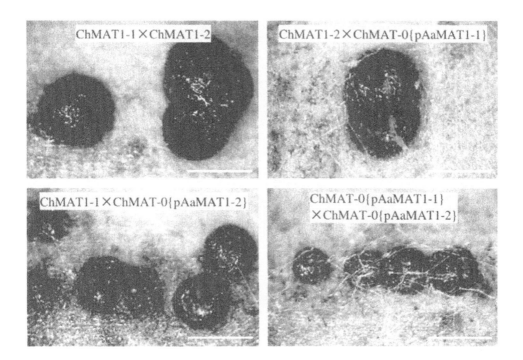

Figure 5. Pseudothecia formation in the crosses. ChMAT1-1=*C. heterostrophus* strain C5 (MAT1-1); ChMAT1-2= strain C4 (MAT1-2); ChMAT-0{pAaMAT1-1}= transformants derived by transformation of MAT null strain C4-41.7 with *A. alternata MAT1-1*; ChMAT-0{pAaMAT1-2}= transformants derived from strain C4-41.7 with *A. alternata MAT1-2*. Pseudothecia formed in the transformant-crosses were indistinguishable from those produced in wild type matings (ChMAT1-1×ChMAT1-2) but the number was smaller than the wild type matings. Bars, 0.5 mm.

3.5.Mating type of field isolates

Determination of mating type of *A. alternata*, *S. turcica*, *Cryphonectria parasitica*, *Nectria haematococca*, *G. fujikuroi*, and *F. oxysporum* field isolates was easily achieved by PCR using MAT-specific primers [30-33]. In the case of the *G. fujikuroi* species complex, isolates from which a 370 bp fragment was amplified with primers Falpha1/Falpha2 were designated

Figure 6.   Ascus and ascospore formation in the crosses.   ChMAT1-1=*C. heterostrophus* strain C5 (MAT1-1); ChMAT1-2= strain C4 (MAT1-2); ChMAT-0{pAaMAT1-1}= transformants derived by transformation of MAT null strain C4-41.7 with *A. alternata MAT1-1*; ChMAT-0{pAaMAT1-2}= transformants derived from strain C4-41.7 with *A. alternata MAT1-2*.   Only a few asci and ascospores were produced in the transformant-crosses.

MAT1-1 and those from which a 260 bp fragment (HMG-box) was amplified with primers FHMG11/FHMG12 were designated MAT1-2 [32]. The conventional mating type minus (–) strains of mating populations A-E of *G. fujikuroi* and a mating type plus (+) strain of mating population F were designated MAT1-1 and mating type + strains of mating populations A-E and a mating type – strain of mating population F were designated MAT1-2 [32]. The effectiveness of the PCR-based mating type assessment for *G. fujikuroi* has recently been confirmed by Steenkamp *et al.* [34]. The mating type of asexual *F. oxysporum* field isolates was also determined by PCR using the same primer sets [31]. No strain failed to show a band with both sets of primers and no strain showed amplification of both products, suggesting that *F. oxysporum* was originally heterothallic. All of the strains of *F. oxysporum* f. sp. *lycopersici* (pathogen causing tomato wilt), carried MAT1-1 [31] except strain MN-60, obtained from Professor H.C. Kistler, University of Minnesota which should be from a different lineage [35]. This suggested two possible explanations of asexuality in *F. oxysporum* f. sp. *lycopersici*: 1) the fungus is a female sterile mutant from the original heterothallic fungus and has become a collection of asexual clones or 2) the fungus still has the ability to reproduce sexually but no mating partner is present in the population at this moment [31].

REFERENCES

1. J.-z. Xiao, A. Ohshima, T. Kamakura, T. Ishiyama and I. Yamaguchi, Mol. Plant-Microbe Interact., 7 (1994) 639.
2. J. E. Hamer, R. J. Howard, F. G. Chumley and B. A. Valent, Science, 239 (1988) 288.
3. J. L. Beckerman and D. J. Ebbole, Mol. Plant-Microbe Interact., 9 (1996) 450.
4. A. Galat, Eur. J. Biochem., 216 (1993) 689.
5. N. J. Talbot, D. J. Ebbole and J. E. Hamer, Plant Cell, 5 (1993) 1575.
6. A. Justesen, S. Somerville, S. Christiansen and H. Giese, Gene, 170 (1996) 131.
7. R. A. Dean, Annu. Rev. Phytopathol., 35 (1997) 211.
8. J. R. Xu and J. E. Hamer, Genes Dev., 10 (1996) 2696.
9. S. Liu and R. A. Dean, Mol. Plant-Microbe Interact., 10 (1997) 1075.
10. W. B. Choi and R. A. Dean, Plant Cell, 9 (1997) 1973.
11. J.-R. Xu, C. J. Staiger and J. E. Hamer, Proc. Natl. Acad. Sci. USA, 95 (1998) 12713.
12. T. K. Mitchell and R. A. Dean, Plant Cell, 7 (1995) 1869.
13. A. Vidal-Cros, F. Viviani, G. Labesse, M. Boccara and M. Gaudry, Eur. J. Biochem., 219 (1994) 986.
14. S. Kang, J. A. Sweigard and B. Valent, Mol. Plant-Microbe Interact., 8 (1995) 939.
15. C.-S. Hwang, M. A. Flaishman and P. E. Kolattukudy, Plant Cell, 7(1995) 183.
16. T. Motoyama, K. Imanishi and I. Yamaguchi, Biosci. Biotech. Biochem., 62 (1998) 564.
17. S. Kagabu and Y. Kurahashi, J. Pesticide Sci., 23 (1998) 145.
18. T. Motoyama, K. Imanishi, T. Kinbara, Y. Kurahashi and I. Yamaguchi, J. Pesticide Sci., 23 (1998) 58.
19. M. Nakasako, T. Motoyama, Y. Kurahashi and I. Yamaguchi, Biochemistry, 37 (1998) 9931.
20. A. Andersson, D. Jordan, G. Schneider and Y. Lindqvist, Structure, 4 (1996) 1161.
21. A. Andersson, D. Jordan, G. Schneider and Y. Lindqvist, FEBS Letters, 400 (1997) 173.
22. J.A. Lucas, Plant Pathology and Plant Pathogens., Blackwell Science, UK, 1998.
23. E. Coppin, R. Debuchy, S. Arnaise and M. Picard, Microbial Mol. Biol. Rev., 61 (1977) 411.
24. B.G. Turgeon, Annu.Rev. Phytopathol., 36 (1998) 115.

25. N.L. Glass, J. Grotelueschen and R.L. Metzenberg, PNAS, 87 (1990) 4912.

26. C. Staben and C. Yanofsky, PNAS, 87 (1990) 4917.

27. R. Debuchy and E. Coppin, Mol. Gen. Genet., 233 (1992) 113.

28. B.G. Turgeon, H. Boelmann, L.M. Ciuffetti, S.K. Christiansen, G. Yang, W. Schafer and O.C. Yoder, Mol. Gen. Genet., 238 (1993) 270.

29. R.L. Metzenberg and N.L. Glass, Bioessays, 12 (1990) 53.

30. T. Arie, S.K. Christiansen, O.C. Yoder and B.G. Turgeon, Fungal Genet. Biol., 21 (1997) 118.

31. T. Arie, I. Kaneko, T, Yoshida, M. Noguchi, Y. Nomura and I. Yamaguchi, Mol. Plant-Microbe Interact., 13 (2000) 1330.

32. T. Arie, T. Yoshida, T. Shimizu, M. Kawabe, K. Yoneyama and I. Yamaguchi, Mycoscience, 40 (1999) 311.

33. S.-H. Yun, T. Arie, I. Kaneko, O.C. Yoder and B.G. Turgeon, Fungal Genet. Biol., 31 (2000) 7.

34. E.T. Steenkamp, B.D. Wingfield, T.A. Coutinho, K.A. Zeller, M.J. Wingfield, F.O. Marasas and J.F. Leslie, Appl. Environ. Microbiol., 66 (2000) 4378.

35. K. O'Donnell, H.C. Kistler, E. Cigelnik and R.C. Ploetz, PNAS, 95 (1998) 2044.

36. S. Kang, F.G. Chumley and B. Valent, Genetics, 138 (1994) 289.

37. G. Singh and A.M. Ashby, Mol. Microbiol., 30 (1998) 799.

Molecular Anatomy of Cellular Systems
I. Endo et al., (editors)
© 2002 Elsevier Science B.V. All rights reserved.

A novel type of Na$^+$/H$^+$ antiporter: its unique characteristics and function

Saori Kosono, Makio Kitada, and Toshiaki Kudo

Laboratory of Microbiology, RIKEN (The Institute of Physical and Chemical Research)
Wako, Saitama 351-0198, JAPAN

A Na$^+$/H$^+$ antiporter, which appears to predominantly contribute to the alkaliphily of *Bacillus halodurans* C-125, was studied in an alkali-sensitive mutant of this strain and a transformant with restored alkaliphily. The Na$^+$/H$^+$ antiporter was found to play a dominant role in pH homeostasis important for alkaliphily in *B. halodurans* C-125. A *Bacillus subtilis* homologue (ShaA) of the Na$^+$/H$^+$ antiporter plays a major role in extrusion of cytotoxic Na$^+$. These antiporters belong to a novel family of cation/H$^+$ antiporters encoded by a cluster of seven genes and are widely distributed in bacteria. The antiporters of this type are involved in a variety of cellular functions including cell development and infection, not only pH homeostasis and Na$^+$ extrusion during vegetative growth.

1. INTRODUCTION

Sodium ions (Na$^+$) are one of the most common types of ions and they sometimes exert stress or have toxic effects on cells [1]. Cells always extrude Na$^+$ actively and maintain relatively low cytoplasmic levels of these ions, but it is not clear whether it is of primary importance to maintain an inwardly directed gradient of Na$^+$ across the cell membrane or to maintain a cytoplasmic Na$^+$ concentration below a certain threshold level. Though Na$^+$ toxicity has been recognized for a long time, the specific cellular targets susceptible to Na$^+$ toxicity have not been identified, except in the case of yeast [2][3]. The nature of Na$^+$ toxicity remains to be elucidated.

Na$^+$ play a primary role in cell bioenergetics as a substitute for H$^+$. It has been reported that some halotolerant bacteria have respiratory Na$^+$ pump systems which contribute to Na$^+$ detoxification and which also may generate a sodium-motive force [4]. It has been mentioned that F-type ATPases capable of using Na$^+$ as coupling ions may function as ATP synthases [4]. Na$^+$ are used as coupling ions for solute transport. The toxicity of Na$^+$ may be related to their ability to substitute for H$^+$, and thus compete with H$^+$. Na$^+$ are not nuisance ions to be eliminated from the cell, and circulation of Na$^+$ may have a role in the cell. K$^+$, another abundant type of monovalent cation, have relatively weak toxicity and are preferentially accumulated inside the cell.

Secondary Na$^+$/H$^+$ antiporters are known to function as a major Na$^+$ extrusion system in all living cells [5][6]. They are driven by a proton-motive force (PMF) generated by respiration or ATP hydrolysis. The Na$^+$/H$^+$ antiporter facilitates influx of H$^+$ in exchange for

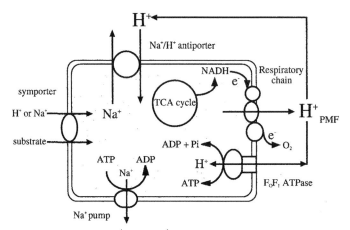

Figure 1 H and Na cycles in bacterial cells

Na, but the ion flux is reversible in mammalian cells [6]. The Na efflux facilitated by the transporter plays an important role in halotolerance and osmotolerance, while the H influx contributes to pH homeostasis [5][6][7]. We have investigated the mechanism of bacterial alkaliphily and have identified a Na/H antiporter which has unique and interesting features compared with other known antiporters. We describe here the characteristics and function of this novel type of Na/H antiporters found in alkaliphilic and non-alkaliphilic bacteria.

2.   Na$^+$/H$^+$ ANTIPORTER RESPONSIBLE FOR THE ALKALIPHILY OF *Bacillus halodurans* C-125

Alkaliphiles grow optimally or very well under pH conditions above pH 9 and usually require sodium ions for their growth [8]. How alkaliphiles maintain a neutral cytoplasmic pH, i.e. pH homeostasis, when exposed to an alkaline environment is one of the most important topics in the study of alkaliphiles. Several different mechanisms have been suggested to play a role in such pH homeostasis [9]. Aono et al. (1999) have recently shown that an acidic teichuronopeptide polymer in the cell wall serves as a barrier to flux of relevant ions and plays a role in pH homeostasis in the facultative alkaliphile *Bacillus halodurans* C-125 [10]. On the other hand, several reports have suggested that Na/H antiporters that operate electrogenically may play a predominant role in pH homeostasis [11][12][13][14][15]. We have used *B. halodurans* C-125 as a model of alkaliphiles, since 1) this strain is a facultative alkaliphile, thus it offers the opportunity to study alkali-sensitive mutants experimentally; 2) it grows well in minimal medium, which is advantageous in selection of auxotrophic markers introduced into mutants of this strain; 3) a host-vector system has been developed; 4) data on the entire genome sequence of both *B. halodurans* C-125 and *B. subtilis* are available, and it is evident that *B. halodurans* C-125 is closely related to *B. subtilis*, thus, a lot of data on *B. subtilis* are available for use as control data in analysis of neutralophiles.

2.1. Isolation and characterization of alkali-sensitive mutants

For use in investigation of the mechanism(s) responsible for the alkaliphily of *B. halodurans* C-125, we isolated several alkali-sensitive mutants which had lost the ability to grow above pH 9.5. Six alkali-sensitive mutants showing no growth at pH 10.5, although each grew well at pH 7.5, were obtained [16]. Only one of the mutants, #38154, was unable to sustain a low internal pH when exposed to an alkaline environment, whereas the other five mutants still retained this ability [16]. We assumed that strain #38154 had lost the function of pH homeostasis, important for alkaliphily, and further studied it.

There is general agreement that the Na^+/H^+ antiporter is driven by the ΔpH (transmembrane proton gradient) component and/or the $\Delta\Psi$ (transmembrane potential) component of the PMF [17][18]. However, alkaline pH is an adverse condition for generating ΔpH, and, under such conditions, only the $\Delta\Psi$ component would contribute to the PMF. Thus, we examined the $\Delta\Psi$-dependent Na^+/H^+ antiport activity by measuring $^{22}Na^+$ efflux or H^+ influx in right-side-out membrane vesicles derived from the parent strain and the mutant, under an imposed $\Delta\Psi$ [19]. $\Delta\Psi$ (interior negative) was created by potassium diffusion in the presence of valinomycin. The imposed $\Delta\Psi$ led to intravesicular acidification of membrane vesicles of the parental strain C-125, but not those of the mutant #38154. Consistent with this result, in response to the imposed $\Delta\Psi$, there was a rapid efflux of $^{22}Na^+$ from membrane vesicles of the parent strain, but not from those of the mutant. These results indicate that the $\Delta\Psi$-dependent Na^+/H^+ antiporter catalyzed uphill H^+ influx coupled with Na^+ efflux when an artificial $\Delta\Psi$ was imposed. Since the transmembrane potential imposed was equal in the case of the vesicles of both the parental and mutant strains, it can be concluded that the $\Delta\Psi$-dependent Na^+/H^+ antiporter in mutant #38154 is unable to function electrogenically at alkaline pH.

2.2. Na^+/H^+ antiporter-coding gene responsible for the alkaliphily

Using the alkali-sensitive mutant #38154 as a host for protoplast transformation, we isolated DNA fragments that restored the alkaliphily in the mutant. A transformant of mutant #38154 harboring pALK2 could grow to the same extent as did the parental strain C-125 over the pH range from 7 to 10.5, whereas mutant #38154 only grew below pH 9 [20]. Consistent with the growth phenotype, the transformant harboring pALK2 showed a restored ability to maintain an internal pH lower than that of the external milieu and it also showed restored Na^+/H^+ antiport activity driven by $\Delta\Psi$ [21]. Nucleotide sequence analysis showed that pALK2 contained parts of two ORFs (ORF1 and ORF3) and one complete ORF (ORF2). In order to identify the DNA region in pALK2 responsible for the complementation, deleted derivatives of pALK2 were constructed and each was tested for the capacity to restore alkaliphilic growth in mutant #38154 [21]. The mutation site was within the region between the *BclI* and *AccI* sites in ORF1, as shown in Figure 2. Direct sequencing of the corresponding region in mutant #38154 revealed a G to A substitution, which resulted in an amino acid substitution from Gly-393 to Arg in the putative ORF1 product [21]. Thus, it can be concluded that ORF1 is responsible for $\Delta\Psi$-dependent Na^+/H^+ antiport activity required for pH homeostasis in alkaline environments, a mechanism contributing to the alkaliphily of *B. halodurans* C-125. This is the first report to show the existence of a Na^+/H^+ antiporter-encoding gene responsible for bacterial alkaliphily. From recent results of analysis of the whole genome sequence of *B. halodurans* C-125, it has been found that the Na^+/H^+ antiporter is encoded within a seven-gene operon [22][23].

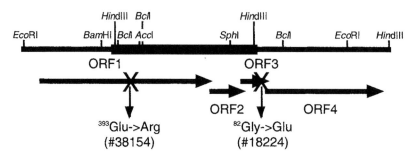

Figure 2 Positions of amino acid substitutions responsible for the alkali-sensitivity of mutants #38154 and #18224. A 5.1 kb EcoRI-HindII fragment from strain C-125 is shown together with ORF1 to ORF4 (arrows) located in the fragment. A HindIII fragment cloned in pALK2 which was effective to restore the alkaliphily of the mutant #38154 is indicated by a thick line.

Another alkali-sensitive mutant, strain #18224, was found to have a mutation resulting in an amino acid substitution in the 82nd residue of the ORF3 product (Figure 2) [24]. Mutant #18224 still retains the ability to control the internal pH, though it shows alkali-sensitive growth [16]. It appears that ORF3 is not involved in Na^+/H^+ antiport itself, but it may be involved in a regulatory process or some other function associated with ion transport.

3. THE *sha* LOCUS OF *Bacillus subtilis*

3.1. A *B. subtilis* homologue of the Na^+/H^+ antiporter responsible for the alkaliphily of *B. halodurans* C-125

From the results of analysis of the whole genome sequence of *B. subtilis*, a homologue (*yufT* [22] or *shaA* [25] or *mrpA* [23]) of the Na^+/H^+ antiporter-coding gene responsible for the alkaliphily of *B. halodurans* C-125 was identified. The ShaA protein shows high similarity to the ORF1 product of *B. halodurans* C-125 (56.8% identity) and its gene is located within a seven-gene cluster.   As mentioned below, this is a common feature of this type of antiporters.

To confirm whether ShaA is responsible for Na^+/H^+ antiport activity and to analyze its role in the neutralophile *B. subtilis*, we disrupted the *shaA* gene and characterized the growth phenotype as a function of Na^+ concentration, pH, and Na^+/H^+ antiport activity [26]. We detected decreased Na^+/H^+ antiport activity in the *shaA* mutant compared with the wild type in assays using right-side-out membrane vesicles upon energization with a transmembrane proton gradient (ΔpH). The *shaA* mutant also showed severe Na^+ sensitivity in its growth. In a Tris-buffered LB-based medium (containing 12 mM endogenous Na^+) at pH 7, the specific growth rate of the *shaA* mutant decreased as the NaCl concentration was increased up to 200 mM, whereas that of the wild type did not change. At pH 8, the *shaA* mutant was more sensitive to Na and no longer grew in the presence of 50 mM NaCl. The growth rate at pH 7 or 8 when the endogenous Na was limited less than 1 mM was identical comparing the wild type and the mutant [26]. Considering that in *E. coli* loss of all three major antiporters causes the same degree of Na sensitivity as that seen in the case of the *shaA* mutant [27], it seems likely that the ShaA antiporter plays a dominant role in the extrusion of cytotoxic Na in *B. subtilis*. The toxic effect of Na on the shaA mutant as measured in terms of growth was diminished in the presence of K, as generally observed with other Na sensitive mutants [28].

Ito et al. (1999) have reported that the products of the *mrp* (identical to *sha*) genes are involved in pH homeostasis and cholate resistance as well as Na resistance [23]. They have suggested that the products of the *mrp* genes are related to both Na - and K -dependent pH homeostasis and may also have some K /H antiport capacity [23]. They have more recently shown that all seven *mrp* genes are required for Na resistance, whereas MrpF (ShaF) can function in cholate resistance independently of the products of the other *mrp* genes [29]. Mutations in any of the *mrp* genes result in a significant increase in *mrp* expression [23][29].

3.2. Role of ShaA in sporulation

B. subtilis cells initiate the formation of dormant endospores when they face growth-limiting and stressful conditions [30]. The primary environmental signal for initiation of sporulation is nutrient depletion. Factors having severe effects on sporulation are often those linked to a decrease in growth rate or growth yield. However, we found a range of NaCl concentration where sporulation of the *shaA* mutant was inhibited without affecting

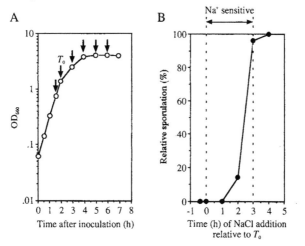

Figure 3. Effect of NaCl addition on sporulation of the *shaA* mutant. A. Growth curve of the *shaA* mutant in 2 x SG medium. 30 mM NaCl was added to the medium at the times indicated with arrows. T0 indicates the end of the exponential growth phase. B. The relative sporulation is shown as the number of spores produced per ml relative to that in the case when grown without added NaCl.

vegetative growth. In a nutrient broth-based sporulation medium (containing 18 mM endogenous Na$^+$), the *shaA* mutant produced $10^5 \sim 10^7$ spores/ml, whereas the wild type produced 10^8 spores/ml [25]. The partial defect in sporulation of the *shaA* mutant was considered to be attributable to the contaminating Na$^+$. The addition of 30 mM NaCl completely blocked the sporulation of the *shaA* mutant (less than 10 spores/ml), though it had little affect on its vegetative growth. Furthermore, sporulation of the *shaA* mutant was still completely blocked when the cells were initially grown in the absence of added NaCl, with 30 mM NaCl added at the end of the exponential growth phase. When NaCl was added within 3 h after the end of the exponential growth phase, it had an inhibitory effect on sporulation, but not after that (Figure 3). Thus, it seems evident that the ShaA antiporter plays a significant role in post-vegetative growth or sporulation, not only vegetative growth. The *shaA* mutant sporulates normally when the Na$^+$ concentration is limited to less than 1 mM [25]. In contrast, K$^+$ has no inhibitory effect, but instead slightly promotes sporulation of the *shaA* mutant.

The period of time when the *shaA* mutant showed Na$^+$ sensitivity in its sporulation was confined to the 3 h period after the end of the exponential growth phase, suggesting that specific targets of Na$^+$ may be expressed in this period. This corresponds to the period of initiation of sporulation, when two key transcription factors, σ^H and phosphorylated SpoOA (SpoOA~P), accumulate and are activated, and then the cells enter the early stage of sporulation [31][32]. The initiation of sporulation is believed to be primarily regulated by the intracellular level of SpoOA~P which controls, sometimes together with σ^H, the expression of early sporulation genes positively or negatively [31]. Since SpoOA~P is generated through an active multicomponent phosphorelay, whether the phosphorelay is activated or not is a point of divergence in the initiation of sporulation. Transcription of *spoOA* is regulated by two

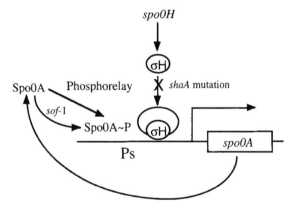

Figure 4. Regulation of *spoOA* expression from the Ps promoter. Transcription of *spoOA* requires the σ^H-containing RNA polymerase and SpoOA~P. The SpoOA proteins are phosphorylated through a phosphorelay system and activate their own transcription in a feedback manner. The *sof-1* mutation can bypass the need for the phosphorelay in the generation of SpoOA~P [34]. The *shaA* mutation affects post-transcriptional regulation of σ^H, but not transcription of *spoOH* and the phosphorelay.

promoters, Pv and Ps. Pv is recognized by a σ^A-containing RNA polymerase (RNAP) during vegetative growth and Ps is recognized by a σ^H-containing RNAP at an early stage of sporulation [33]. Transcription of *spo0A* from the Ps promoter also requires its own product Spo0A~P and this allows activation of the phosphorelay in a positive feedback manner (Figure 4). To determine whether the Na -sensitive steps are before or after activation of the phosphorelay, we first examined the expression of *spo0A* from the Ps promoter in the *shaA* mutant.

In the absence of 30 mM NaCl, the rate of *spo0A* (Ps) induction was lower in the *shaA* mutant compared with that in the wild type [25]. Moreover, *spo0A* (Ps) induction was almost completely blocked in the *shaA* mutant in the presence of 30 mM NaCl and also by addition of 30 mM NaCl at the end of the exponential growth phase [25]. There results suggest that the *shaA* mutation blocks sporulation before activation of the phosphorelay by affecting the generation of Spo0A~P and/or the active σ^H-containing RNAP. However, a mutation bypassing the phosphorelay (*sof*-1 mutation [34]) did not restore *spo0A* (Ps) induction in the *shaA* background, suggesting that the primary defect in sporulation of the *shaA* mutant is not directly related to phosphorylation of Spo0A [25]. Thus, it was considered that a loss of σ^H function, a decrease in the level of σ^H and/or a decrease in the transcription activity of the σ^H-containing RNAP, occurs in the *shaA* mutant. We then assayed the level of σ^H protein by Western blot analysis using anti-σ^H antibody and found that the accumulation of σ^H was impaired in the *shaA* mutant in the presence of NaCl. Transcription of *spo0H* encoding σ^H was not affected, indicating that this impaired accumulation of σ^H occurred at the post-transcriptional level [25].

σ^H is an alternative sigma factor that governs transcription in stationary phase [35]. It has been clearly demonstrated that the level of σ^H is regulated mainly by a post-transcriptional mechanism(s) [36], but the detailed features of the mechanism(s) remain to be elucidated. The ATP-dependent protease Lon [37], the regulatory ATPase ClpX [37][38], holoenzyme formation with RNA polymerase [39], and pH [40] have been suggested to be involved in the mechanism. The observation that the *shaA* mutation affected post-transcriptional regulation of σ^H leads us to speculate that a reaction(s) or interaction(s) sensitive to Na may be involved in the post-transcriptional regulation of σ^H.

4. Sha HOMOLOGUES IN OTHER BACTERIA

The ORF1 product of *B. halodurans* C-125 and ShaA of *B. subtilis* have no similarity to other known Na /H antiporters encoded by a single gene. Recent reports have demonstrated that they belong to a novel family of monovalent cation/H antiporters usually encoded by a cluster of seven genes. This type of antiporter is widely distributed among bacteria, including *Shinorhizobium* (*Rhizobium*) *meliloti* (*pha*) [41], *Staphylococcus aureus* (*mnh*) [42], alkaliphilic *B. pseudofirmus* OF4, and others annotated in the genome database. DNA regions homologous to the *sha* cluster have been found also in the genome of radiation-resistant *Deinococcus radiodurans*, the purple nonsulfur bacterium *Rhodobacter capsulatus*, and thermophilic *Thermotoga maritima*, but not in *E. coli*.

Table 1 shows a comparison of the Na /H antiporter homologues in *B. subtilis*, *B. halodurans* C-125, *S. meliloti*, and *S. aureus*. There are very close similarities in terms of gene organization and protein sizes. The *pha* locus of *S. meliloti* was identified as a gene region required for infection of leguminous plants by these bacteria to establish nitrogen-fixing symbiosis [41]. It is suggested that the *pha* locus may be a region encoding a K^+/H^+ antiporter since *pha* mutants show sensitivity to K^+, but not Na^+, in their growth and they are deficient in diethanolamine-induced K^+ efflux [41]. This is consistent with the observation that the *mrp* (or *sha*) gene product may have $Na^+(K^+)/H^+$ antiport activity, as mentioned above [23]. *pha* mutations were found to be distributed over a cluster of seven genes [41].

Table 1 A comparison of Sha proteins and their homologues.

	shaA	shaB	shaC	shaD	shaE	shaF	shaG

Strain	*Bacillus subtilis*	Alkaliphilic *Bacillus halodurans* C-125	*Sinorhizobium meliloti*	*Staphylococcus aureus*
Protein	ShaA (774 aa)	BH1319 (ORF1) (804 aa)	PhaA (725 aa)	MnhA (801 aa)
	ShaB (143 aa)	BH1318 (ORF2) (146 aa)	PhaB (257 aa)	MnhB (142 aa)
	ShaC (113 aa)	BH1317 (ORF3) (112 aa)	PhaC (115 aa)	MnhC (113 aa)
	ShaD (493 aa)	BH1316 (ORF4) (493 aa)	PhaD (547 aa)	MnhD (498 aa)
	ShaE (158 aa)	BH1315 (158 aa)	PhaE (161 aa)	MnhE (159 aa)
	ShaF (94 aa)	BH1314 (95 aa)	PhaF (92 aa)	MnhF (97 aa)
	ShaG (124 aa)	BH1313 (117 aa)	PhaG (120 aa)	MnhG (118 aa)
Function	Na^+/H^+ antiporter Major role in Na^+ excretion	Na^+/H^+ antiporter pH homeostasis under alkaline conditions	K^+/H^+ antiporter Involved in infection	Na^+/H^+ antiporter

a) This table was prepared based on data presented in Table 1 of the report by Kitada et al. (2000) [43].

b) Part of *B. halodurans* C-125 data was exported from the website of ExtremoBase [44].

The *mnh* locus, a region encoding a Na^+/H^+ antiporter of *S. aureus*, was cloned using an *E. coli* mutant lacking the three major Na^+/H^+ antiporters as a host [42]. Deleted derivatives lacking *mnhA*, *mnhD*, or *mnhG* did not complement the NaCl sensitive growth of the *E. coli* mutant host [42]. It was shown that the seven *mrp* (or *sha*) genes are transcribed as an operon [23]. Evidence suggests that all seven genes are required for the ion transport function. We have shown that mutations in the first gene result in loss of Na^+/H^+ antiport activity in *B. halodurans* C-125 and *B. subtilis*, thus the first gene is necessary for, but not the sole contributor to, the ion transport function. On the other hand, as mentioned above, the third gene of the corresponding cluster in *B. halodurans* C-125 seems not to be involved in ion transport directly [16] and MrpF (or ShaF) of *B. subtilis* seems to function in cholate resistance independently of the other genes [28]. These antiporter-coding genes may have complex functions and the role of each of the seven genes in the transport function remains to be elucidated.

Interestingly, several Sha proteins show similarity to subunits of respiratory-coupled

NADH dehydrogenase complexes. The ShaA and ShaD homologues are related to NuoL and NuoM of *E. coli* NADH:ubiquinone oxidoreductase (complex I), respectively [43]. ShaB, ShaC, and ShaE also show similarity to other redox proteins [28]. However, Na⁺ extrusion by Mrp (or Sha) or Mnh is severely inhibited by a proton conductor, CCCP [23][29][42] [Kosono, unpublished results], and primary pump activity has not been detected in the case of these antiporters.

5. CONCLUSIONS

The Sha antiporter and its homologues show unique and interesting features. They are encoded by a cluster of seven genes, whereas all other known cation antiporters are encoded by a single gene. They show secondary antiport activity in spite of appearing to function as primary pumps. They also show variation in their characteristics and functions, in terms of their capacity for Na⁺, K⁺, and cholate transport, their involvement in sporulation, infection, and cholate resistance, as well as pH homeostasis and Na⁺ resistance. Fine control of cytoplasmic ion concentrations, including H⁺, Na⁺, and K⁺ levels, may be important for cellular processes or functions such as sporulation and infection, and antiporters belonging to this family seem to play a predominant role in such ionic regulation. Through genome sequencing, homologues have been found in other bacteria, and it is of interest to clarify the function of each corresponding cluster.

REFERENCES

1. E. Padan and T.A. Krulwich, in Bacterial stress response, (G. Storz and R. Hengge-Aronis, eds.), ASM press, Washington, D.C., pp. 117, (2000).
2. H.-U. Glaser, D. Thomas, R. Gaxiola, F. Montrichard, Y. Surdin-Kerjan and R. Serrano, EMBO J., 12 (1993) 3105.
3. J.R. Murgufa, J.M. Belles and R. Serrano, J. Biol. Chem., 271 (1996) 29029.
4. P. Dimroth, Biochim. Biophys. Acta, 1318 (1997) 11.
5. E. Padan and S. Schuldiner, Biochim. Biophys. Acta, 1185 (1994) 129.
6. J. Orlowski and S. Grinstein, J. Biol. Chem., 272 (1997) 22373.
7. T.A. Krulwich, J. Cheng and A.A. Guffanti, J. Exp. Biol., 196 (1994) 457.
8. K. Horikoshi, Microbiol. Mol. Biol. Rev., 63 (1999) 735.
9. T.A. Krulwich, M. Ito, R. Gilmour and A.A. Guffanti, Extremophiles, 1 (1997) 163.
10. R. Aono, M. Ito and T. Machida, J. Bacteriol., 181 (1999) 6600.
11. A.A. Guffanti and T.A. Krulwich, J. Biol. Chem., 255 (1980) 7391.
12. M. Kitada, A.A. Guffanti and T.A. Krulwich, J. Bacteriol., 152 (1982) 1096.
13. T.A. Krulwich, A.A. Guffanti, R.F. Bornstein and J. Hoffstein, J. Biol. Chem., 257 (1982) 1885.
14. T.A. Krulwich, J. Membr. Biol., 89 (1986) 113.
15. M. Kitada, K. Onda and K. Horikoshi, J. Bacteriol., 171 (1989) 1879.
16. M. Hashimoto, T. Hamamoto, M. Kitada, M. Hino, T. Kudo and K. Horikoshi, Biosci. Biotech. Biochem., 58 (1994) 2090.

17. M. Bassilana, E. Damiano and G. Leblanc, Biochemistry 23 (1984) 1015.
18. M. Bassilana, E. Damiano and G. Leblanc, Biochemistry 23 (1984) 5288.
19. M. Kitada, M. Hashimoto, T. Kudo and K. Horikoshi, J. Bacteriol., 176 (1994) 6464.
20. T. Kudo, M. Hino, M. Kitada and K. Horikoshi, J. Bacteriol., 172 (1990) 7282.
21. T. Hamamoto, M. Hashimoto, M. Hino, M. Kitada, Y. Seto, T. Kudo and K. Horikoshi, Mol. Microbiol., 14 (1994) 939.
22. B. Oudega, G. Koningstein, L. Rodrigues, M.S. Ramon, H. Hilbert, A. Dusterhoft, T.M. Pohl and T. Weizenegger, Microbiology, 143 (1997) 2769.
23. M. Ito, A.A. Guffanti, B. Oudega and T.A. Krulwich, J. Bacteriol., 181 (1999) 2394.
24. Y. Seto, M. Hashimoto, R. Usami, T. Hamamoto, T. Kudo and K. Horikoshi, Biosci. Biotech. Biochem., 59 (1995) 1364.
25. S. Kosono, Y. Ohashi, F. Kawamura, M. Kitada and T. Kudo, J. Bacteriol., 182 (2000)898.
26. S. Kosono, S. Morotomi, M. Kitada and T. Kudo, Biochim. Biophys. Acta, 1409 (1999) 171.
27. T. Sakuma, N. Yamada, H. Saito, T. Kakegawa and H. Kobayashi, Biochim. Biophys. Acta, 1363 (1998) 231.
28. M. Harel-Bronstein, P. Dibrov, Y. Olami, E. Pinner, S. Schuldiner and E. Padan, J. Biol. Chem., 270 (1995) 3816.
29. M. Ito, A.A. Guffanti, W. Wang and T.A. Krulwich, J. Bacteriol., 182 (2000) 5663.
30. A. L. Sonenshein, in Bacterial stress response, (G. Storz and R. Hengge-Aronis, eds.), ASM press, Washington, D.C., pp. 199, (2000).
31. J.A. Hoch, Annu. Rev. Microbiol., 47 (1993) 441.
32. A.D. Grossman, Annu. Rev. Genet., 29 (1995) 477.
33. T. Chibazakura, F. Kawamura and H. Takahashi, J. Bacteriol., 173 (1991) 2625.
34. F. Kawamura and H. Saito, Mol. Gen. Genet., 192 (1983) 330.
35. W.G. Haldenwang, Microbiol. Rev., 59 (1995) 1.
36. J. Healy, J. Weir, I. Smith and R. Losick, Mol. Microbiol., 5 (1991) 477.
37. J. Liu, W.M. Cosby and P. Zuber, Mol. Microbiol., 33 (1999) 415.
38. J. Liu and P. Zuber, Mol. Microbiol., 37 (2000) 885.
39. Y. Ohashi, K. Sugimaru, H. Nanamiya, T. Sebata, K. Asai, H. Yoshikawa and F. Kawamura, Gene, 229 (1999) 117.
40. W.M. Cosby and P. Zuber, J. Bacteriol., 179 (1997) 6778.
41. P. Putnoky, A. Kereszt, T. Nakamura, G. Endre, E. Grosskopf, P. Kiss and A. Kondorosi, Mol. Microbiol., 28 (1998) 1091.
42. T. Hiramatsu, K. Kodama, T. Kuroda, T. Mizushima and T. Tsuchiya, J. Bacteriol., 180 (1998) 6642.
43. M. Kitada, S. Kosono and T. Kudo, Extremophiles, 4 (2000) 253.
44. H. Takami, K. Nakasone, Y. Takaki, G. Maeno, R. Sasaki, N. Masui, F. Fuji, C. Hirama, Y. Nakamura, N. Ogasawara, S. Kuhara and K. Horikoshi, Nuc. Acids Res., 28 (2000) 4317.

PART II
NUCLEAR FUNCTIONS

Molecular Anatomy of Cellular Systems
I. Endo et al., (editors)

Genetic analysis of the genes involved in mitosis in fission yeast *Schizosaccharomyces pombe*

Yukinobu Nakaseko[a, b] and Mitsuhiro Yanagida[a, b]

Department of Gene Mechanisms, Graduate School of Biostudies[a], and Department of Biophysics, Graduate School of Science[b], Kyoto University, Kitashirakawa-Oiwakecho, Sakyo-ku, Kyoto 606-8502, Japan

Eukaryotic cells undergo mitosis to transmit genetic information to daughter cells and many genes are involved in this process. For identification and characterization of the genes that are required for mitosis, we have been using fission yeast *Schizosaccharomyces pombe* as a model system. Both isolation of the mutants defective in mitosis and construction of the mutants by recombinant DNA technique are the one of the powerful approaches to identify and characterize such genes at molecular level. Systematic isolation of a series of mitotic mutants and analysis of genes by reverse genetics have revealed a number of molecular networks in regulation of mitosis, and most of the genes identified were evolutionary conserved among the eukaryotic cells.

1.   STRATEGY FOR IDENTIFICATION OF THE FACTORS INVOLVED IN MITOSIS IN FISSION YEAST

For identification of the factors involved in regulation of mitosis, genetical approach is very powerful in yeast system. Although fission yeast *Schizosaccharomyces pombe* has relatively small genome size (14Mb) as for eukaryote, nevertheless retains all basic mechanical components such as mitotic spindle and regulatory factors such as MPF (maturation promoting factor or M-phase promoting factor) (1 - 4). And also, most of the genes involved in regulation or construction of such mitotic apparatus are evolutionary conserved among the eukaryotic cells. Analysis of a series of *cdc* (cell division cycle) genes was one of the most successful examples as a genetical approach in order to identify cell cycle regulators highly conserved among eukaryote (5). We have started systematic isolation of mutants defective in mitosis, mainly focused on the behavior of chromosomes during mitosis. The mutations in genes involved in mitosis are expected to cause defective mutant phenotype in mitotic stage specific manner. Such mitotic mutants were classified with their phenotypes to assume in which stage the wild type genes have essential function (Figure 1).

To investigate the mitotic phenotype of these mutants at high resolution, microscopic technique, mainly fluorescent microscopic technique is one of the most powerful and indispensable techniques. Development of various DNA staining dyes and fluorescent probes combined with epitope tagging technique for detection of a specific protein have made it possible to visualize what is occurring or defective in mutant cells at molecular level.

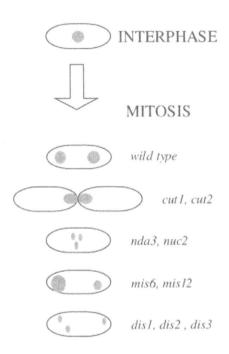

Figure. 1 Schematic representation of mitosis of fission yeast wild type and mutant cells.

1.1.Start from mutant isolation

We have constructed several sets of mutant library in which each mutant shows temperature sensitive growth (Table 1). Then phenotype of each mutant under restrictive temperature was analyzed. The morphological change of chromatin region stained with DAPI is initial criteria for such large-scale screening. In some cases, other criteria such as cell elongation, drug sensitivity or minichromosome instability was combined with this morphological information. Each mutant were classified and named, such as *cut* (cell untimely torn), *dis* (defective in sister chromatid disjoining), *nda* (nuclear division arrest) and *nuc* (6 - 11). *mis* (mini-chromosome instability) mutants have been identified as mutants that lose minichromosome at high frequency at the permissive or semi-restrictive temperature, in addition to showing temperature sensitive growth (12).

Among these mutants in the same mutant phenotype group, some gene products form a protein complex to function. For example, Cut1 and Cut2 form a large complex (13). Cut4 and Cut9 are the subunits of APC/cyclosome that catalyzes polyubiquitination of cyclin and Cut2 as an E3 ubiquitin ligase (14). Cut3 and Cut14 are the subunits of condensin which is essential protein complex required for chromosome condensation during mitosis (15). Nda2 and Nda3 are α- and β-tubulin, respectively (16, 17). Nda1 and Nda4 are the component of MCM (minichromosome maintenance) protein complex that is essential for initiation of DNA replication (18).

Table 1 Genes identified from *cut, dis, mis, nda, nuc* mutants in fission yeast

gene	gene product	function
cut1+	separin	cleavage of Rad21
cut2+	securin	sister chromatid separation
cut3+	condensin subunit	chromosome condensation
cut4+	APC/cyclosome subunit	polyubiquitination of cyclin and Cut2
cut5+	BRCT motif protein	DNA synthesis checkpoint
cut6+	acetyl CoA carboxylase	equal size nuclear division
cut7+	motor protein	spindle microtubule formation
cut8+		proteasome localization
cut9+	APC/cyclosome subunit	polyubiquitination of cyclin and Cut2
cut11+		spindle pole body component (78)
cut12+		spindle pole body component (79)
cut14+	condensin subunit	chromosome condensation
cut15+	importin α	nuclear import, chromosome condensation
cut20+	APC/cyclosome subunit	polyubiquitination of cyclin and Cut2
cut23+	APC/cyclosome subunit	polyubiquitination of cyclin and Cut2
dis1+	ch-TOG related protein	microtubule binding protein
dis2+	protein phosphatase	exit from mitosis
dis3+		component of the RCC1-Ran pathway and exosome
mis1+	Polδ subunit (*cdc1+*)	DNA replication
mis2+	RRN motif protein	
mis3+	KH motif protein	ribosome RNA biogenesis, initiation of cell growth
mis4+	adherin	chromosome cohesion
mis5+	MCM subunit	initiation of DNA synthesis
mis6+	kinetochore protein	equal segregation of sister chromatids
mis10+	Polδ subunit (*cdc6+*)	DNA replication
mis11+	RNA splicing factor (*prp2+*)	RNA splicing
mis12+	kinetochore protein	equal segregation of sister chromatids
nda1+	MCM subunit	initiation of DNA synthesis
nda2+	α-tubulin	component of microtubules
nda3+	β-tubulin	component of microtubules
nda4+	MCM subunit	initiation of DNA synthesis
nuc1+	RNA polymerase I	organization of nucleolus, transcription
nuc2+	APC/cyclosome subunit	polyubiquitination of cyclin and Cut2

1.2.Start from gene cloning

Isolation of mitotic mutant is powerful strategy to identify the factors involved in mitosis. However, sometimes it is difficult to assume the phenotype of their defect. And also, as the rapid progress of whole genome DNA sequencing project, enormous amount of information can be obtained to identify the gene of interest at a nucleotide sequence level. In these case, we can start from gene cloning to construct mutants by reverse genetics. For example, this approach is applicable if homologous gene have been shown to be involved in mitosis in different organisms or microsequencing data is available from purified, unknown proteins. It starts from cloning of DNA fragment containing the genes of interest. Using these DNA fragment, the target gene of chromosomal copy can be either deleted (gene disruption) or replaced with mutagenized version *in vitro*. The genes analyzed by this approach include type 2A protein phosphatases (19) and their regulatory subunits (20), ppe1 protein phosphatase (21), type2B protein phosphatase (22), γ-tubulin (23), the subunits of condensin complex (24), the subunits of cohesin complex (25) and SpCENP-A (26).

1.3.Microscopic technique

Microscopic analysis is essential for identification and characterization of mitotic mutants. Especially, fluorescent microscopic technique is very powerful with the rapid development of wide variety of fluorescent probes and high sensitive camera systems. In fission yeast cells, initial studies with DAPI staining technique has revealed the temporal order of morphological changes of nuclear chromatin during mitosis (10). This analysis has made possible to screen a large amount of mutants with its high-resolution image and relatively rapid staining procedure compared with previous staining technique such as Giemsa staining. Indirect immuno-fluorescence microscopic technique gave more information when combined with DAPI staining. Staining of microtubules and Sad1 (spindle pole body component) are routinely used for analysis of behavior and integrity of spindle microtubules (27). By FISH (fluorescence in situ hybridization) technique, the specific DNA region of interest like centromeric or telomeric DNA can be visualized (28, 29). Also, using the mixture of DNA probes spanning on the specific chromosome "arm" regions at appropriate intervals, the level of chromosome condensation during mitosis can be visualized by FISH analysis. Introduction of Green Fluorescent Protein (GFP) system has made it possible to analyze the temporal localization or morphological change of target proteins in living cells (30, 31). Analysis of Sad1-GFP fusion protein revealed the three different kinetic phases of spindle microtubule dynamics from prophase, metaphase to anaphase (32). As an extended technique of GFP system, lacI-GFP fusion protein system enabled to visualize the behavior of the centromeric DNA of the kinetochore in living cells (33).

2. GENES IDENTIFIED FROM MITOTIC MUTANTS

2.1.*cut* mutants

cut mutants show an uncoordination of nuclear division and cytokinesis at the restrictive temperature. These mutants undergo cytokinesis in the absence of nuclear division, producing the cells with *cut* (for *c*ell *u*ntimely *t*orn) phenotype. The idea of screening for this class of mutants was originated from the analysis of *top2* mutant that is defective in the type II DNA topoisomerase (34). In this mutant, spindle microtubules were formed and pulled the chromosomes in mitosis, but failed to segregate chromosomes to the poles. Resultant undivided chromosomes were 'cut' during cytokinesis (6, 8, 11).

cut1 and *cut2*

In *cut1* and *cut2* mutants, chromatin was extensively stretched without separation of sister chromatids at the restrictive temperature (35). Also multiple rounds of the cell cycle progression in the absence of chromosome separation leads to polyploid and multiple SPB formation when cytokinesis is blocked. These results indicate that Cut1 and Cut2 are required for chromosome separation. Cut1 localizes in the cytoplasm during interphase and moves to the spindle and spindle pole upon the entry into mitotic prophase (36). Cut2 concentrates along the short spindle in metaphase (37). The amount of Cut1 is constant throughout the cell cycle, in contrast to the rapid degradation of Cut2 at the onset of anaphase. Cut1 and Cut2 form large protein complexes (13). This complex formation has been shown to be required for the onset of normal anaphase. This physical interaction may explain that the temperature sensitivity of *cut2* mutant is suppressed with an elevated level of Cut1. This is also consistent with that *cut2* mutant failed to form Cut1/Cut2 complex at the restrictive temperature. Cut1 shows homology to the budding yeast Esp1p/Separin at its carboxy terminus. Esp1p has been shown to be a component of proteolysis of Scc1p that is a component of the cohesion complex. Scc1 is thought to link sister chromatids prior to initiation of anaphase (38).

Cut2 belongs to a conserved gene family called securin which includes the budding yeast Pds1 and human PTTG (Pituitary tumor-transforming gene) (39, 40). Cut2/securin is degraded before anaphase and the anaphase does not occur in the absence of Cut2/securin destruction (37). Moreover, this destruction is dependent on its destruction sequence (40). Like cyclin, securin is a target of APC (anaphase promoting complex)/cyclosome. Cut2/securin is polyubiquitinated in the presence of APC and degraded by proteasome like mitotic cyclin Cdc13. Human securin/PTTG is an oncogene. High level expression of PTTG takes place in some tumor cell lines and exhibits transforming activity in NIH 3T3 cells. This transforming activity may results from chromosome aberration caused by incorrect sister chromatids separation.

cut3 and *cut14*

Both *cut3* and *cut14* mutants failed to condense chromosomes but small portions of chromosomes can separate along the spindle microtubules. By FISH analysis, the contraction of the chromosome arms during mitosis has been shown to be defective. These mutant chromosomes may not be rigid enough to be pulled toward the spindle poles (41). Cut3 and Cut14 are the subunits of condensin complex. Condensin is evolutionary conserved, essential protein complex for chromosome condensation during mitosis. Cut3 and Cut14 show homology each other as a member of gene family called SMC (structural maintenance of chromosome) family, including Smc2p and Smc4p of the budding yeast (41). Although condensin complex is composed of five protein components, Cut3 and Cut14 can form a complex *in vivo* (15). The purified Cut3/Cut14 complex posses single stranded DNA renaturation activity. This DNA renaturation activity is approximately 70-fold more efficient than that of E. coli RecA and heat sensitive in *cut3* mutant cells. Cut3/Cut14 complex may be involved in higer order DNA coiling process in chromatin condensation with its DNA renaturation activity (15). The rest of subunits of condensin, named Cnd1, Cnd2 and Cnd3 have been clone and analyzed (see the chapter for condensin).

cut4 and cut9

cut4 and *cut9* mutants show similar phenotype at the restrictive temperature. These mutants show block or delay in the progression from metaphase to anaphase (14, 42). Highly condensed chromosomes and short spindles are accumulated in these mutants, followed by cytokinesis with cutting undivided nuclei. Both Cut4 and Cut9 are the subunits of APC/cyclosome that has mitosis specific E3 ubiquitin ligase activity (14). In these mutant cells, APC/cyclosome formation was affected and polyubiquitination of Cdc13, a mitotic B-type cyclin, was decreased. Cut4 is a evolutionary conserved protein including the budding yeast Apc1p, mouse Tsg24 and *Aspergillus nidulans* BimE (14). Cut9 is also evolutionary conserved protein including the budding yeast Apc6p/Cdc16p. The growth defect and dissociation of APC/cyclosome complex in *cut4* mutant are suppressed by introduction of $cgs2^+/pde1^+$ (encoding a cAMP phosphodiesterase) or $pka1^+$ (encoding a cyclic AMP dependent protein kinase catalytic subunit) on multicopy plasmid. These results indicate that APC/cyclosome formation and activation are under the negative regulation by cAMP/PKA pathway (14, 43).

cut5

$cut5^+$ is identical to $rad4^+$, which was previously identified as a radiation sensitive mutant gene (44). *cut5* mutants are defective in initiation and/or elongation of DNA replication but allow mitosis and cell division at a restrictive temperature. *cut5* mutant cells enter mitosis followed by cytokinesis even in the absence of completion of DNA synthesis. However, when DNA is damaged in mutant cells, cell division is arrested. Thus, Cut5 is an essential component of the DNA replication checkpoint system but not the DNA damage checkpoint. Cut5 has a repeat motif, which resembles the N-terminal repeat domain of proto-oncogene product Ect2. From the screening of the gene physically binds to Cut5, $crb2^+$ was identified (45). Crb2 is required for checkpoint arrests induced by irradiation and polymerase mutations. The carboxyl terminal region of Crb2 shows homology to yeast Rad9p, human p53BP1 and BRCA1. Cut5 and Crb2 interact with Chk1, which is essential protein kinase on checkpoint pathway, in a yeast two-hybrid assay. Moreover, ectopic expression of Chk1 suppresses the phenotypes of *cut5* and *crb2* mutants. These results suggest Cut5, Crb2, and Chk1 form an essential protein complex for checkpoint pathway. Crb2 is phosphorylated at T215 in response to DNA damage *in vivo*. Nonphosphorylatable mutant at T215 site remains arrested state after DNA damage, even after repairing of DNA. And cdc2 kinase can phosphorylate T215 site *in vitro*. These results suggest that phosphorylation of Crb2 by cdc2 kinase is required for reentering the cell cycle after DNA damage-induced checkpoint arrest (46).

cut6

cut6 mutant shows defective nuclear division resulting in unequal size of daughter nuclei. Interestingly, $cut6^+$ encodes acetyl CoA carboxylase, a key enzyme in fatty acid synthesis (47). Also, $lsd1^+$, which mutant shows similar phenotype to that of *cut6* mutant, encodes fatty acid synthetase. Moreover, cerulenin, an inhibitor of fatty acid synthesis, causes unequal size of nuclei during nuclear division in wild type cells. These results show the requirement of fatty acid for normal nuclear division.

cut7

In *cut7* mutant cells, a novel V-shaped spindle microtubules are observed at the restrictive temperature instead of formation of normal mitotic spindles (48). This V-shaped microtubules are always associated with the chromosomes. In this mutant, interdigitation in spindle microtubule does not appear to take place. Instead, two unconnected half spindle microtubules are formed and fails to chromosome separation. Cut7 belongs to a gene family encoding kinesin heavy chain, including bimC of *Aspergillus nidulans* and Eg5 of *Xenopus laevis*. These results indicate that essentiality of kinesin motor molecule for mitotic spindle formation during mitosis.

cut8

cut8 mutant shows chromosome hypercondensation and short-spindle formation in the absence of sister chromatid separation at the restrictive temperature, followed by cytokinesis with cutting undivided nuclei. Cut8 is an evolutionary conserved protein including the budding yeast Dbf8p/Sts1p (49). Homologous genes have been also found in *Candida albicans* and *Botrytis cinerea*. Disruption of *cut8*$^+$ leads to temperature sensitive growth and its phenotype was similar to those found in the original mutant. *cut8* mutant shows synthetic lethality with mutations impaired in APC/cyclosome function (*cut4*, *cut9* and *slp1*), proteasome (*mts2* and *mts3*) and a type 1 protein phosphatase (*dis2*) (50). These genetic interaction are consistent with the idea that Cut8 is implicated in promoting anaphase through interaction with ubiquitin-dependent proteolysis pathway. From microscopic analysis of living cells using GFP tagged proteins, degradation of both Cdc13 and Cut2 has been shown to be delayed in *cut8* mutant. Cut8 is localized to the nucleus and nuclear periphery. This localization pattern is highly similar to those of 26S proteasome subunits. Localization of 26S proteasome is affected in *cut8* mutant. Only faint nuclear localization of the components of proteasome (Mts2 and Mts3) was observed in *cut8* mutant that is contrast to their nuclear localization in wild type cells. These results suggest Cut8 facilitates ubiquitin dependent proteolysis by recruiting 26S proteasome to the nucleus (50).

cut15

cut15 mutant shows loss of chromosome condensation at the restrictive temperature. This phenotype is very similar to those of *top2* and *cut14* mutants, type II DNA topoisomerase and a condensin subunit mutant, respectively. Interestingly, Cut15 is a member of importin α, which is essential protein for NLS (nuclear localization signal) dependent nuclear localization with importin β (51). And also, purified Cut15 retained the identical activity to that of mammalian importin α in a nuclear import assay. However, in *cut15* mutant at the restrictive temperature, NLS proteins were imported into the nucleus as wild type cells. Cut15 is essential for mitotic chromosome condensation, but its activity in nuclear transport might be dispensable.

cut20 and cut23

Screening of the mutant which shows altered sedimentation profiles of APC/cyclosome among the newly identified *cut* mutants identified *cut20* and *cut23* mutants (52). Both Cut20 and Cut23 are subunits of APC/cyclosome. Cut20 and Cut23 are homologue of the budding yeast APC/cyclosome components Apc4p and Apc8p, respectively. Interestingly, similar to *cut4* mutant, growth defect of *cut20* mutant is suppressed by introduction of $cgs2^+/pde1^+$ (encoding a cAMP phosphodiesterase) or $pka1^+$ (encoding a cyclic AMP dependent protein kinase catalytic subunit) on multicopy plasmids.

2.2. *dis* mutants

dis mutants are isolated as cold sensitive mutants that are defective in mitosis. In these mutants, chromosomes are highly condensed and spindle microtubules elongate, however, sister chromatids are never separated (9). Shape of elongated spindle microtubule is aberrant and their staining pattern is fainter in the central region compared with those of wild type cells. As the terminal phenotype, three highly condensed chromosomes and V-shaped spindle microtubules are observed. In these arrested cells, H1 kinase activity increases and Cdc13, a mitotic B-type cyclin are accumulated. These results suggest *dis* mutant cells are arrested at the late stage of mitosis. Three loci, $dis1^+$, $dis2^+$ and $dis3^+$, are identified as mutant genes that show *dis* phenotype. Phenotype of these three mutants are highly similar and any pair of these three mutation cause synthetic lethality. These results indicate the existence of overlapping function required for sister chromatid separation during mitosis in these three genes.

dis1

Both *dis1* mutant and disruptant cells show cold sensitive growth and *dis* phenotype. Dis1 is a microtubule binding protein (53, 54). Dis1 colocalized with both cytoplasmic microtubules and spindle microtubules in the cell. Dis1 has 6 major phosphorylation sites and at least 3 of them clustered at its carboxy terminal region are phosphorylated by cdc2 kinase *in vitro*. These phosphorylation sites have been shown to be phosphorylated *in vivo* as well (53). The biological significance of these phosphorylation is not known, but one possibility is that the affinity between microtubule and Dis1 may be regulated in mitosis specific manner. From living cell analysis using GFP tagged $dis1^+$ gene, Dis1 has been shown to localize at the kinetochore region in metaphase to anaphase (55). The interaction between Dis1 and the centromeric DNA in the kinetochore region has also been detected by CHIP (chromatin immunoprecipitation) analysis. Thus, Dis1 localizes to the kinetochore and/or the kinetochore microtubules in mitosis specific manner. Dis1 may stabilize kinetochore microtubules in metaphase by interacting with the ends of microtubules and so counteract the action of microtubule destabilizing factors during anaphase.

Dis1 shares homology to a gene family containing human ch-TOG (colonic and hepatic tumor over-expressed gene), frog XMAP215, fly msps (minispindles), budding yeast Stu2p (suppressor of tubulin) and nematode ZYG-9. The homologous region consists of either two or four copies of large repeated sequence. All of these proteins colocalized with microtubules or centrosomes/spindle pole bodies in the cell. This homologous region may share same function among the genes in this family, possibly regulating dynamic behavior of microtubular architecture in the cell. Dis1 also has small repeated motif called HEAT motif (56). HEAT repeats are found in various chromosome-associated proteins, including frog XCAP-D2 and XCAP-G (condensin), the budding yeast Scc2p (cohesin), TBP-associated

TIP120 protein, the budding yeast Mot1p (SWI2/SNF2 family) and the fission yeast Mis4 (adherin). HEAT repeats motifs may have a common function involved in chromosome dynamics or microtubule function.

dis2

$dis2^+$ encodes a catalytic subunit of type 1 protein phosphatase (PP1) (57). Its cold sensitive mutant $dis2$-11 shows dis phenotype indistinguishable to those of $dis1$ or $dis3$ mutants. Dis2 is highly conserved protein in eukaryotic cells. Originally, PP1 was identified as a key regulatory molecule in glycogen metabolic pathway. Analysis of $dis2$ mutant has revealed new functional aspect of PP1. PP1 is essential molecule in mitosis as well as metabolic pathway. Dis2 has phosphorylation site of cdc2 kinase at its carboxy terminus. This sequence, thr-pro-pro-arg, is also highly conserved among a number of eukaryotic PP1. Phosphorylation of this site by cdc2 kinase in $vitro$ decreases PP1 phosphatase activity. These results suggest dis2 phosphatase activity is negatively regulated during mitosis by cdc2 kinase (58).

As high dosage suppressor of $dis2$-11 cold sensitive growth, $sds21^+$, $sds22^+$ and $sds23^+$ have been identified. $Sds21^+$ encodes another catalytic subunit of PP1 of fission yeast (57). The presence of $sds21^+$ explains the reason why disruption of $dis2^+$ does not cause lethality to the cell. Disruption of both $dis2^+$ and $sds21^+$ cause lethal and cells are arrested at mitosis. $sds22^+$ encodes a regulatory subunit of PP1. Sds22 binds to both Dis2 and Sds21 (59). Binding of Sds22 to PP1 change substrate specificity when using S1 peptides and phosphorylase as substrates (60). Truncated version of $sds22^+$ in amino terminal region causes temperature sensitive growth and disruption of $sds22^+$ cause lethality. Both truncated mutant cells and disruptant cells are arrested at mitosis. From these results, Sds22 is thought to be a mitosis specific regulator of PP1.

Sds23 is a novel protein genetically interacting with a component of APC/cyclosome as well as PP1 (61). High dosage of $sds23^+$ suppresses the temperature sensitivity of $nuc2$ and $cut9$ mutants. Both Nuc2 and Cut9 are component of APC/cyclosome. Deletion of $sds23^+$ causes both temperature (36°C) and cold (22°C) sensitive growth. In this disruptant cells, the progression of anaphase and cytokinesis is retarded and cell shape is aberrant. These defects are suppressed by high dosage of $nuc2^+$, $cut9^+$ or $sds21^+$, but not by $dis2^+$. Interestingly, Sds23 is neither component of Dis2 phosphatase nor the APC/cyclosome complex in $vivo$. Although the molecular function of $sds23^+$ is not clear, these results indicate that the presence of close functional relationship between Sds23, PP1 and APC/cyclosome during mitosis.

dis3

$dis3$ mutant shows dis phenotype indistinguishable to those of $dis1$ or $dis2$ mutants (62). Interestingly, increased gene dosage of $dis3^+$ reverses the Ts+ phenotype (called bypass of $wee1$ suppression) of a $cdc25$ $wee1$ double mutants, as does increased gene dosage of $dis2^+$. $dis3^+$ was also identified as a multicopy suppressor of disruptant of $ppe1^+$. $ppe1$ encodes a protein phosphatase that shows homology to both type1 and type 2A protein phosphatases. Dis3 may also function through protein phosphorylation/dephosphorylation reaction similar to Dis2. Dis3 was found to be structurally and functionally conserved from yeast to mammals. In the budding yeast and mammalian cell, Dis3 has been shown to be a component of the RCC1-Ran pathway (63). Dis3 binds directly to Ran and enhances the GEF activity of RCC1.

Dis3 shares homology with the budding yeast Ssd1p or Srk1p, which suppress the defect of Sit4p protein phosphatase and the Bcy1p regulatory subunit of cyclic AMP-dependent protein kinase, respectively (62). Sts5 in fission yeast also shares homology to Dis3 (64). Sts5 is involved in polarized growth and functionally interacts with Ppe1 serine/threonine phosphatase, protein kinase C, and Pyp1 tyrosine phosphatase. Interestingly, Dis3 was also identified as a component of exosome (65). Exosome is a highly conserved eukaryotic RNA processing complex containing multiple 3'-->5' exoribonucleases. Exosome of S. cerevisiae consists of the five essential proteins Rrp4p, Rrp41p, Rrp42p, Rrp43p, and Rrp44p. In these subunits, Rnp44p is identical to Dis3p and shows homology to RNase II, while Rrp41p, Rrp42p, and Rrp43p are related to RNase PH.

2.3. *mis* mutants

mis mutants are isolated as mutants showing temperature sensitive growth and which lose minichromosome at high frequency at the permissive or semi-restrictive temperature (*mis* for minichromosome instability). These mutants are isolated from a temperature sensitive growth mutant collection that has minichromosome. Minichromosome stability was visualized by colony colour assay using *ade6*$^+$ gene complementation system. All the *mis* genes identified are essential for viability. Some of them are involved in nucleic acid metabolism rather than direct interaction to chromosome separation or segregation function. Mis1, Mis5 and Mis10 are essential for DNA replication (12). *mis1*$^+$ and *mis10*$^+$ are identical to *cdc1*$^+$ and *cdc6*$^+$, respectively. Both *mis1*$^+$/*cdc1*$^+$ and *mis10*$^+$/*cdc6*$^+$ encode subunits of DNA polymerase δ (66, 67). *mis5*$^+$ encodes a subunit of MCM (minichromosome maintenance) protein complex, which is required for early stage of DNA replication (12). *mis11*$^+$ is identical to *prp2*$^+$, and encodes homologue of human pre-mRNA splicing factor U2AF (12). Mis2 is an RNA binding protein that has an RNA binding motif called RRM (68).

mis3

Mis3 is a conserved RNA binding protein that has KH motif. Mis3 is essential for ribosome RNA biogenesis and initiation of cell growth after nutritional starvation (68). In *mis3* mutant at the restrictive temperature, the level of 18S ribosomal RNA is greatly diminished and assembly of 40S ribosome is undetectable. Mis3 shows genetic interaction with RNA processing proteins and cell cycle regulators. *mis3* mutant is synthetic lethal with either *dsk1*$^+$ disruptant or *dis3* mutants whose gene products are involved in RNA metabolism. On the other hand, *mis3* mutation is suppressed by mutation in *dis2*$^+$, which encodes a catalytic subunit of type 1 protein phosphatase. *mis3* mutant also shows hypersensitivity with hydroxyurea and this sensitivity is increased when combined with *cds1*$^+$ deletion. Mis3 may be required for the coupling between growth and cell cycle with supporting S-phase checkpoint.

mis4

Mis4 is an evolutionary conserved protein and essential for viability (69). Mis4 functions as a chromosome cohesion molecule that is independent of cohesin. In *mis4* mutant, chromosome hypercondensation occurs, but sister chromatids are separated in metaphase-arrested cells. And also, premature sister chromatids separation occurs in interphase after DNA replication in *mis4* mutant. *mis4* mutant shows low frequency of the *cut* phenotype in addition to the *mis* phenotype. *mis4* mutant also shows hypersensitive to hydroxyurea and to ultraviolet irradiation at the permissive temperature, and shows synthetic lethality when combined to DNA ligase mutant. Mis4 ensures faithful anaphase chromosome separation and is required in S phase.

mis6

mis6 mutant shows missegregation of regular chromosome in addition to instability of minichromosomes. In this mutant, unequal segregation of sister chromatids was frequently observed at the restrictive temperature (70) Interestingly, Mis6 acts before or at the onset of S phase. Mitotic missegregation defects in *mis6* mutants require the passage of G1/S at the restrictive temperature. Mis6 localizes to the kinetochore throughout the cell cycle. Mis6 has been shown to bind central region of the centromeric DNA in the kinetochore region. Consistently, kinetochore chromatin structure is altered in *mis6* mutant cells. In micrococcal nuclease digestion experiments, smeared pattern at the centromere region which are observed in wild type cells was abolished, and was converted to regular size of nucleosome ladder. Thus Mis6 is an essential kinetochore component in fission yeast.

mis12

mis12 mutant shows missegregation of regular chromosome in addition to instability of minichromosomes, like *mis6* mutant (71). In *mis12* mutant , extension of metaphase spindle was observed. This may be incorrect spindle morphogenesis due to impaired sister centromeres or force unbalance between pulling and pushing at sister kinetochores. This spindle abnormality was observed in *mis6* mutant as well. Mis12 localizes to the kinetochore throughout the cell cycle like Mis6. ChIP analyses have shown that Mis12 binds to the inner centromere region like Mis6. However, these two protein were not detected as a protein complex in immuno-precipitation experiment. And interaction between Mis12 and the centromeric DNA was not affected in *mis6* mutant. Also, Mis6 could interact with the centromeric DNA in the absence of Mis12. Although Mis12 is an essential kinetochore component in fission yeast like Mis6, Mis6 and Mis12 appeared to interact independently to the centromeric DNA in the kinetochore.

2.4. *nda* mutants

nda mutants were isolated as mutants showing cold sensitive growth and defect in nuclear division. These mutants arrested at the restrictive temperature showing elongated cells with single undivided nucleus (nuclear division arrest). Interestingly, Both Nda1 and Nda4 are subunits of MCM (minichromosome maintenance) protein complex that required for early stage of DNA replication (18, 72). Nda2 and Nda3 are α1- and β-tubulin, respectively (16, 17). At permissive temperature, *nda2* mutant shows hypersensitivity to a drug, thiabendazole (TBZ) which is an inhibitor of microtubule polymerization. By complementation screening of cold sensitivity of *nda2* mutant with multi copy plasmid based genomic DNA library, another α-tubulin gene in fission yeast, named α2-tubulin gene, was identified as a dosage dependent suppressor (16).

2.5. *nuc* mutants

nuc1

nuc1 mutant shows a defective nuclear phenotype with an aberrant nucleolus at the restrictive temperature (73). Interestingly, *nuc1*⁺ encodes a large subunit of RNA polymerase I. Nuc1 has been shown to localize at the nucleolus. This nucleolar localization was dependent on the presence of DNA topoisomerases I and II. These results suggest RNA polymerase I is required for the formation of the nucleolus as its major component, and DNA topoisomerases are required for the functional organization of nucleolus.

nuc2

nuc2 mutant shows highly condensed chromosomes with short uniform spindle microtubules at the restrictive temperature (7). Nuc2 has been shown to be a component of APC/cyclosome, together with Cut4 and Cut9 (14). This is consistent with that *nuc2* mutants are arrested at mid mitosis at the restrictive temperature. In addition to mitotic defect, *nuc2* mutant showed sterile phenotype (74). This sterility is due to defect of DNA replication under nutrient starvation. And also, septation takes place in the absence of chromosome separation at the semi-restrictive temperature. Nuc2 may negatively regulate septum formation. This idea is consistent with that overproduction of *nuc2*⁺ blocks septation. An interesting 34 amino acid long sequence motif called snap helix or TPR (tetratricopeptide repeat) was initially identified in this protein (75). This motif has been found in a various proteins including rat protein phosphatase 5 (PP5), the budding yeast cell division control protein Cdc16p and Cdc23p, the budding yeast glucose repression mediator protein Ssn6p, mouse FK506 binding protein FKBP5 and the budding yeast peroxisome assembly protein PAS10p. This motif is thought to be required for recognition of protein- protein interaction.

3.   GENES IDENTIFIED FROM GENE CLONING

3.1. Protein phosphatases and their regulatory proteins

Fission yeast has two genes, *ppa1*⁺ and *ppa2*⁺, encoding type 2A protein phosphatase (PP2A) (19). Although disruption of both gene leads to lethal phenotype as those of two PP1 genes, *dis2*⁺ and *sds21*⁺, the cell cycle defect caused by decreasing phosphatase activity are very different. Introduction of cold sensitive mutations in the homologous region found in PP1 (*dis2-11*) leads to premature mitosis, in contrast to the production of chromosome nondisjunction in *dis2-11* mutant. Consistently, PP1 and PP2A genes are not functionally exchangeable. Thus, the activity of PP1 and PP2A have distinct, but essential roles in growth and cell cycle control (19).

Deletion of *ppa2*⁺ gene leads to premature entry of mitosis with the reduction in cell size (76). This semi-wee phenotype is enhanced in the presence of a phosphatase inhibitor, okadaic acid. *ppa2*⁺ interacts genetically with the cell regulators *cdc25*⁺ and *wee1*⁺. Disruptant of *ppa2*⁺ is lethal when combined with *wee1-50* and partially suppresses the temperature sensitive lethality of *cdc25-22* mutation. These results suggest that Ppa2 negatively regulates the entry into mitosis through the regulation of cdc2 kinase activity by controlling tyrosine phosphorylation/dephosphorylation by Wee1 and Cdc25 (76).

ppe1⁺ was identified as a gene showing homology to both *dis2*⁺ (PP1) and *ppa2*⁺ (PP2A) phosphatases in similar extent (39% and 53% identity, respectively) (21). Disruption of *ppe1*⁺

cause cold sensitive growth and abnormal cell shape. Interestingly, the cold sensitive growth was suppressed by high dosage of either two PP2A genes, $ppa1^+$ or $ppa2^+$ but not by PP1 gene, $dis2^+$. And also, $dis3^+$ and $pck1^+$ (protein kinase C-like kinase) suppressed the cold sensitivity of $ppe1^+$ disruptant cells. These genetic interaction suggest that $ppe1^+$ may be involved in regulation of both cell morphogenesis and mitosis through protein phoshorylation/dephosphorylation.

$ppb1^+$ encodes a type 2B protein phosphatase (PP2B) (22). The $ppb1^+$ disruptant cells show pleiotropic defects including in cytokinesis, mating, transport, nuclear and spindle pole body positioning, and cell shape. Multiply septated phenotype in $ppb1^+$ disruptant cells are also observed in wild type cells treated with Cyclosporin A, a immunosuppresant. Also, Disruptant cells was not affected with cyclosporin A. These results are consistent with that PP2B is a target of cyclosporin A in mammalian cells.

$paa1^+$ and $pab1^+$ encode the regulatory A subunit (PR65) and B subunit (PR55) of the type 2A protein phosphatase, respectively (20). The $paa1^+$ gene was essential for growth and disruptant cells show abnormal distributions of microtubule and actin. Also $paa1^+$ disruptant cells could not form a polarized cell shape. The $pab1^+$ disruptant cells show both cold and temperature sensitivity. $pab1^+$ disruptant cells also show delayed cytokinesis, abnormal cell shape and lost the polar distributions of actin and microtubules. Also, the $pab1^+$ disruptant cells were defective in cell wall synthesis and sporulation. From these results, these two fission yeast PP2A regulatory subunits are implicated in cell morphogenesis, probably through regulation of the cytoskeletal network and cell wall synthesis.

3.2.cohesin

Cohesin is an evolutionary conserved multiprotein complex required for sister chromatid cohesion (77). Budding yeast cohesin is composed of Smc1p, Smc3p, Scc1p/Mcd1p, and Scc3p, whose defects induce premature sister chromatid separation. Among these subunits, only $rad21^+$, a homologue of Scc1p had been identified as radiation sensitive mutant gene in fission yeast. Based on their homology, $psm1^+$, $psm3^+$ and $psc3^+$, homologue of Smc1p, Smc3p, and Scc3p respectively, were cloned in fission yeast (25). In S. pombe, the majority of cohesin was located in the nucleus and seemed to bind to chromatin throughout the cell cycle. This is the major difference of the fission yeast cohesion molecules compared with those of vertebrate and the budding yeast. And also, chromatin immunoprecipitation experiment shows that the fission yeast cohesin subunits are enriched in broad centromere regions and that the level of centromere-associated Rad21 did not change from metaphase to anaphase. Similar to the budding yeast Scc1p/Mcd1p, Rad21 contains similar cleavage sites to those of Scc1p/Mcd1p and a small fraction of Rad21 is cleaved specifically at this site in anaphase. This cleavage is essential for progression of mitosis. Interestingly, loading Rad21 on chromatin is dependent on the presence of Mis4, which is another sister chromatid cohesion protein.

3.3.condensin

The condensin protein complex is required for chromosome condensation in mitosis (77). This complex consists of five proteins, including SMC proteins Cut3 and Cut14 (24). Condensin complex was purified and the rest of three non-SMC proteins, Cnd1, Cnd2 and Cnd3, were identified by microsequencing. These three genes are highly conserved in primary sequence. Gene disruption experiment showed deletion of these genes leads to phenotype indistinguishable from that of *cut3* or *cut14* mutants. GFP-tagged proteins of these complexes were observed to alter localization in cell cycle dependent manner. During mitosis, Cut3 enriched in the nucleus and in cytoplasm in other stages of the cell cycle. This stage specific localization is depending on mitosis specific phosphorylation of Cut3 by cdc2 kinase. cdc2 kinase may regulate nuclear accumulation of condensin complex required for chromosome condensation by phosphorylation of Cut3 during mitosis (24).

3.4.CENP-A

CENP-A is a human protein associate with the kinetochore throughout the cell cycle. SpCENP-A was cloned by polymerase chain reaction that is based on its homology to mammalian CENP-A (26). SpCENP-A has been shown to be required for formation of centromere specific chromatin structure that associates with equal chromosome segregation during mitosis. SpCENP-A deleted cells shows unequal nuclear division during mitosis, followed by cytokinesis producing aneuploid cells at high frequency. In this disruptant cell, centromere specific chromatin structure observed in wild type cells is also altered. ChIP analyses have shown that SpCENP-A binds to the inner centromere region. Interestingly, this centromere localization of SpCENP-A depends on Mis6 but not on Mis12, both are essential centromere proteins in fission yeast.

3.5. γ-tubulin

γ-tubulin belongs to the tubulin family protein that is conserved in eukaryotic cells. This molecule localizes at the centrosome and required for microtubule nucleation from the centrosome. *gtb1*⁺ gene was isolated by its homology to *Aspergillus nidulans* γ-tubulin gene (23). Gtb1 localized at the spindle pole body throughout the cell cycle. Also Gtb1 was associated with microtubule organizing center which appears in telophase and cytokinesis. This localization suggests that γ-tubulin is required for microtubule nucleation in fission yeast as well. *gtb1*⁺ gene is essential for viability and disruptant cells show condensed, undivided chromosomes with aberrant spindle structure. These results indicate that γ-tubulin has evolutionary conserved not only in structural but also functional point of view.

REFERENCES

1. I.M. Hagan and J.S. Hyams, J. Cell Sci., 89 (1988) 343.
2. C.F. Robinow and J.S. Hyams in Molecular biology of the fission yeast: General cytology of fission yeast, (A. Nasim eds.), Academic Press, San Diego, pp. 273, (1989).
3. T. Mizukami, W.I. Chang, I. Garkavtsev, N. Kaplan, D. Lombardi, T. Matsumoto, O. Niwa, Kounosu, M. Yanagida, T.G. Marr and et al., Cell, 73 (1993) 121.
4. S. Su and M. Yanagida in The molecular and celluar biology of the yeast *Saccharomyces*: Mitosis and cytokinesis in the fission yeast, *Schizosaccharomyces pombe*, (J.N. Strathern, J.R. Broach eds.).
5. P. Nurse, Nature, 344 (1990) 503.
6. T. Hirano, S. Funahashi, T. Uemura and M. Yanagida, EMBO J., 5 (1986) 2973.
7. T. Hirano, Y. Hiraoka and M. Yanagida, J. Cell Biol., 106 (1988) 1171.

8. I. Samejima, T. Matsumoto, Y. Nakaseko, D. Beach and M. Yanagida, J. Cell Sci., 105 (1993) 135.

9. H. Ohkura, Y. Adachi, N. Kinoshita, O. Niwa, T. Toda and M. Yanagida, EMBO J., 7 (1988) 1465.

10. T. Toda, M. Yamamoto and M. Yanagida, J. Cell Sci., 52 (1981) 271.

11. M. Yanagida, Trends Cell Biol., 8 (1998) 144.

12. K. Takahashi, H. Yamada and M. Yanagida, Mol. Biol. Cell, 5 (1994) 1145.

13. H. Funabiki, K. Kumada and M. Yanagida, EMBO J., 15 (1996) 6617.

14. Y.M. Yamashita, Y. Nakaseko, I. Samejima, K. Kumada, H. Yamada, D. Michaelson and M. Yanagida, Nature, 384 (1996) 276.

15. Sutani and M. Yanagida, Nature, 388 (1997) 798.

16. T. Toda, Y. Adachi, Y. Hiraoka and M. Yanagida, Cell, 37 (1984) 233.

17. Y. Hiraoka, T. Toda and M. Yanagida, Cell, 39 (1984) 349.

18. T. Okishio, Y. Adachi and M. Yanagida, J. Cell Sci., 109 (1996) 319.

19. N. Kinoshita, H. Ohkura and M. Yanagida, Cell, 63 (1990) 405.

20. K. Kinoshita, T. Nemoto, K. Nabeshima, H. Kondoh, H. Niwa and M. Yanagida, Genes Cells, 1 (1996) 29.

21. M. Shimanuki, N. Kinoshita, H. Ohkura, T. Yoshida, T. Toda and M. Yanagida, Mol. Biol. Cell, 4 (1993) 303.

22. T. Yoshida, T. Toda and M. Yanagida, J. Cell Sci., 107 (1994) 1725.

23. T. Horio, S. Uzawa, M.K. Jung, B.R. Oakley, K. Tanaka and M. Yanagida, J. Cell Sci., 99 (1991) 693.

24. T. Sutani, T. Yuasa, T. Tomonaga, N. Dohmae, K. Takio and M. Yanagida, Genes Dev., 13 (1999) 2271.

25. T. Tomonaga, K. Nagao, Y. Kawasaki, K. Furuya, A. Murakami, J. Morishita, T. Yuasa, T. Sutani, S.E. Kearsey, F. Uhlmann, K. Nasmyth and M. Yanagida, Genes Dev., 14 (2000) 2757.

26. K. Takahashi, E.S. Chen and M. Yanagida, Science, 288 (2000) 2215.

27. I. Hagan and M. Yanagida, J. Cell Biol., 129 (1995) 1033.

28. S. Uzawa and M. Yanagida, J. Cell Sci., 101 (1992) 267.

29. H. Funabiki, I. Hagan, S. Uzawa and M. Yanagida, J. Cell Biol., 121 (1993) 961.

30. A.B. Cubitt, R. Heim, S.R. Adams, A.E. Boyd, L.A. Gross and R.Y. Tsien, Trends Biochem. Sci., 20 (1995) 448.

31. H. Tatebe, G. Goshima, K. Takeda, T. Nakagawa, K. Kinoshita and M. Yanagida, Micron 32 (2001) 67.

32. K. Nabeshima, T. Nakagawa, A.F. Straight, A. Murray, Y. Chikashige, Y.M. Yamashita, Y. Hiraoka and M. Yanagida, Mol. Biol. Cell, 9 (1998) 3211.

33. C.C. Robinett, A. Straight, G. Li, C. Willhelm, G. Sudlow, A. Murray and A.S. Belmont, J. Cell Biol., 135 (1996) 1685.

34. T. Uemura and M. Yanagida, EMBO J., 3 (1984) 1737.

35. S. Uzawa, I. Samejima, T. Hirano, K. Tanaka and M. Yanagida, Cell, 62 (1990) 913.

36. K. Kumada, T. Nakamura, K. Nagao, H. Funabiki, T. Nakagawa and M. Yanagida, Curr. Biol., 8 (1998) 633.

37. H. Funabiki, H. Yamano, K. Kumada, K. Nagao, T. Hunt and M. Yanagida, Nature, 381 (1996) 438.

38. F. Uhlmann, D. Wernic, M.A. Poupart, E.V. Koonin and K. Nasmyth, Cell, 103 (2000) 375.
39. M. Yanagida, Genes Cells, 5 (2000) 1.
40. H. Funabiki, H. Yamano, K. Nagao, H. Tanaka, H. Yasuda, T. Hunt and M. Yanagida, EMBO J., 16 (1997) 5977.
41. Y. Saka, T. Sutani, Y. Yamashita, S. Saitoh, M. Takeuchi, Y. Nakaseko and M. Yanagida, EMBO J., 13 (1994) 4938.
42. I. Samejima and M. Yanagida, J. Cell Biol., 127 (1994) 1655.
43. H. Yamada, K. Kumada and M. Yanagida, J. Cell Sci., 110 (1997) 1793.
44. Y. Saka and M. Yanagida, Cell, 74 (1993) 383
45. Y. Saka, F. Esashi, T. Matsusaka, S. Mochida and M. Yanagida, Genes Dev., 11 (1997) 3387.
46. F. Esashi and M. Yanagida, Mol. Cell, 4 (1999) 167.
47. S. Saitoh, K. Takahashi, K. Nabeshima, Y. Yamashita, Y. Nakaseko, A. Hirata and M. Yanagida, J. Cell Biol., 134 (1996) 949.
48. I. Hagan and M. Yanagida, Nature, 356 (1992) 74.
49. I. Samejima and M. Yanagida, Mol. Cell Biol., 14 (1994) 6361.
50. H. Tatebe, G. Goshima, K. Takeda, T. Nakagawa, K. Kinoshita and M. Yanagida, Micron, 32 (2001) 67.
51. T. Matsusaka, N. Imamoto, Y. Yoneda and M. Yanagida, Curr. Biol., 8 (1998) 1031.
52. Y.M. Yamashita, Y. Nakaseko, K. Kumada, T. Nakagawa and M. Yanagida, Genes Cells, 4 (1999) 445.
53. K. Nabeshima, H. Kurooka, M. Takeuchi, K. Kinoshita, Y. Nakaseko and M. Yanagida, Genes Dev., 9 (1995) 1572.
54. Y. Nakaseko, K. Nabeshima, K. Kinoshita and M. Yanagida, Genes Cells, 1 (1996) 633.
55. Y. Nakaseko, G. Goshima, J. Morishita and M. Yanagida, Curr. Biol., 11 (2001) 537.
56. A.F. Neuwald and T. Hirano, Genome Res., 10 (2000) 1445.
57. H. Ohkura, N. Kinoshita, S. Miyatani, T. Toda and M. Yanagida, Cell, 57 (1989) 997.
58. H. Yamano, K. Ishii and M. Yanagida, EMBO J., 13 (1994) 5310.
59. H. Ohkura and M. Yanagida, Cell, 64 (1991) 149.
60. E.M. Stone, H. Yamano, N. Kinoshita and M. Yanagida, Curr. Biol., 3 (1993) 13.
61. K. Ishii, K. Kumada, T. Toda and M. Yanagida, EMBO J., 15 (1996) 6629.
62. N. Kinoshita, M. Goebl and M. Yanagida, Mol. Cell Biol., 11 (1991) 5839.
63. E. Noguchi, N. Hayashi, Y. Azuma, T. Seki, M. Nakamura, N. Nakashima, M. Yanagida, X. He, U. Mueller, S. Sazer and T. Nishimoto, EMBO J., 15 (1996) 5595.
64. T. Toda, H. Niwa, T. Nemoto, S. Dhut, M. Eddison, T. Matsusaka, M. Yanagida and D. Hirata, J. Cell Sci., 109 (1996) 2331.
65. P. Mitchell, E. Petfalski, A. Shevchenko, M. Mann and D. Tollervey, Cell, 91 (1997) 457.
66. S.A. MacNeill, S. Moreno, N. Reynolds, P. Nurse and P.A. Fantes, EMBO J., 15 (1996) 4613.
67. Y. Iino and M. Yamamoto, Mol. Gen. Genet., 254 (1997) 93.
68. H. Kondoh, T. Yuasa and M. Yanagida, Genes Cells, 5 (2000) 525.
69. K. Furuya, K. Takahashi and M. Yanagida, Genes Dev., 12 (1998) 3408.
70. S. Saitoh, K. Takahashi and M. Yanagida, Cell, 90 (1997) 131.
71. G. Goshima, S. Saitoh and M. Yanagida, Genes Dev., 13 (1999) 1664.

72. S. Miyake, N. Okishio, I. Samejima, Y. Hiraoka, T. Toda, I. Saitoh and M. Yanagida, Mol. Biol. Cell, 4 (1993) 1003.
73. T. Hirano, G. Konoha, T. Toda and M. Yanagida, J. Cell Biol., 108 (1989) 243.
74. K. Kumada, S. Su, M. Yanagida and T. Toda, J. Cell Sci., 108 (1995) 895.
75. T. Hirano, N. Kinoshita, K. Morikawa and M. Yanagida, Cell, 60 (1990) 319.
76. N. Kinoshita, H. Yamano, H. Niwa, T. Yoshida and M. Yanagida, Genes Dev., 7 (1993) 1059.
77. T. Hirano, Genes Dev., 13 (1999) 11.
78. R.R. West, E.V. Vaisberg, R. Ding, P. Nurse and J.R. McIntosh, Mol. Biol. Cell., 9 (1998) 2839.
79. A.J. Bridge, M. Morphew, R. Bartlett and I.M. Hagan, Genes Dev., 12 (1998) 927.

Molecular Anatomy of Cellular Systems
I. Endo et al., (editors)
© 2002 Elsevier Science B.V. All rights reserved.

Intergenomic transcriptional interplays between plastid as a cyanobacterial symbiont and nucleus

Hideo Takahashi and Kan Tanaka

Institute of Molecular and Cellular Biosciences, The University of Tokyo,Bunkyo-ku, Tokyo 113-0032, Japan

Plastids, plant organelles including chloroplasts, are considered to have originated from an endosymbiotic event of ancient cyanobacteria, the primeval inventor of the oxygen-generating photosyntems, in other eukaryotic cells. The vestiges of cyanobacterial genetic traits are found in both plastid and nuclear genomes. Significant numbers of original cyanobacterial genes evolutionarily disappeared from the plastid genome of extant plant cells; some have lost ever because of the dispensability, and the others have translocated onto the nuclear genome presumably for the regulatory reasons. One of the obstacles to unveil the coordinated gene expression between the two genomic systems, plastid and nucleus, was absence of the genetic information about the sigma subunit, a key factor of the plastid-encoded RNA polymerase (PEP). Actually all of the genes encoding for the multi-subunit core enzyme are found in the plastid genome, but the sigma factor gene is not. By scrutinizing the epigraphs depicted in the common sequences of sigma factors in cyanobacteria, we have successfully identified nuclear genes (*sig*) encoding for plastid sigma factors. This strategy was first adopted for unicellular red algae, *Cyanidium caldarium* RK-1, and then for three higher plants; two of typical dicotyledonous, *Arabidopsis thaliana* and *Nicotiana tabacum*, and one of monocotyledonous, *Orysa sativa*. Open reading frames found in the cDNA clones of these nuclear genes indicate that the N-terminal regions of the gene products had amino acid sequences typical to the plastid-targeting transit peptides. Furthermore, a transient expression assay of GFP fusions in *Arabidopsis* protoplasts showed that the N-termini of these *sig* gene products functioned as chloroplast-targeting signals. The *sigA*- or *sigB*-promoter fused with a *uidA* reporter in the transgenic *Arabidopsis*, was similarly activated at various tissues of the plants, such as cotyledons, hypocotyls, rosette leaves, sepals and siliques, but not at roots, seed, or other flower organs. Promoters including those from *Cyanidium*, *Arabidopsis*, and *Nicotiana* were repeatedly activated under continuous light, somewhat similar to endogenous rhythms. An *Arabidopsis* mutant (*abc1*) having a pale-green leaf phenotype presumably by the impaired *sigB* (= *sig2*) function was isolated as a T-DNA insertion clone. This result provides direct evidence that a nuclear-derived prokaryotic-like SigB protein plays a critical role in the coordination of the two genomes for plastid development.

1. PLANT PLASTIDS AND CYANOBACTERIA

Plastids are plant-specific organelles, possessing a small eubacterial-type genome. Here we discuss how plastids are considered to have originated from an endosymbiotic event of ancient cyanobacteria in other primitive eukaryotic cells.

1.1. Endosymbiotic origin of plant plastids

Plastid is a generic name of divergent plant organelles, such as chloroplasts in green leaves, etioplasts in dark grown seedlings and amyloplasts in storage cells of cereal grains, all of which have ingeniously developed from undifferentiated proplastids in the meristemic primordia. Like cyanobacteria, chloroplasts of the plant cells are specialized to carry out oxygen-producing photosynthesis (the oxygenic photosynthesis) with two photosystems where water is the electron donor and oxygen is the ultimate oxidation product (1). Actually, all of the basic photosynthetic apparatuses of plant cells are localized in chloroplasts. And several lines of evidence support a hypothesis that plant plastids have derived from endosymbiotic events of ancient cyanobacteria in other eukaryote-like cells. In fact, the processes of the oxygen-generating photosynthesis in cyanobacteria and plant chloroplast are essentially identical.

The key photosynthetic system of chloroplasts has originated from an invention by cyanobacteria that succeeded in creating an oxygen-generating photosystem (the second photosystem) as well as harnessing the perilous light energy, although chloroplasts are very popular as a prominent holder of the oxygen-generating photosynthetic system. The elaboration of the second photosystem by ancient cyanobacteria made it possible to use universally available water as the electron donor for autotrophic carbon dioxide assimilation. Thus, cyanobacteria should be regarded as an admirable inventor of the oxygen-generating photosynthesis, by which extant higher plants flourish as a generous supplier of nutritions for rapacious and parasitic animals on the earth.

1.2. Plastid genome and nuclear-encoded plastid genes

Chloroplasts and also other plastids of plant cells contain their own genomes as multi-copies of a circular double-stranded DNA. Typical plastidic DNA found in higher plants are around 150 kilo-base pairs (kb), on which up to 150 genes encoding rRNA, tRNA and proteins for photosynthesis, transcription, translation, and some other functions are found. The genetic system presumed by the sequence information is a prokaryotic type, which is highly similar to that of cyanobacteria, supporting the cyanobacterial origin of plant plastids (2,3).

The genome size of higher plant plastids is only one-tenth or less of that of extant cyanobacterial genomes. Thus, the number of genes is far less that required for all possible structure and function of higher plant plastids, especially for photosynthesis and plastidic differentiation. Constituents of genes in plastid genomes of various plant cell lineages indicate that the genes absent in plastid genomes were either lost or translocated into other genome (presumably into the nucleus). Since the genetic information in the plastid genome is apparently insufficient for plastid functions, the remaining portions must be supplied from the nuclear genes, which have evolutionarily been translocated from the plastid genome. Actually most macromolecular apparatuses in the plastids, such as ribosome and photosynthetic centers, are composites of both nuclear and plastid gene products.

The chimeric or mosaic constituents of plastid apparatuses are considered as the resultant of translocation of specific genes from plastids to nuclear genome during the evolutionary process of plastids. Thus, genes from both plastid and nuclear genomes are involved in the construction of functionally and structurally divergent plastids. Although a number of nuclear-encoded gene products are translocated into plastids to function as the integral part, to date we do not know the overall features of nuclear genes involvement for the plastid formation and functional differentiation (4).

1.3. Transcription machineries in cyanobacteria and plastids

The basic structure of transcription machinery in cyanobacteria is an RNA polymerase holoenzyme, consisting of the multi-subunit core enzyme and sigma factor. The latter protein confers to the core RNA polymerase to the ability to select specific promoter sequences and to initiate transcription from the specific site. The core enzyme of *E. coli* which is able to elongate RNA chains by incorporating nucleotide triphosphates complementary to the one strand of DNA template, consists of heteromeric tetramer from two molecules of α and each one of β and β' subunits encoded by *rpoA*, *rpoB*, and *rpoC*, respectively (5). The RNA polymerase of cyanobacteria is essentially the same basic constituent as that of *E. coli*, except of the β' protein is split to β' and γ subunit (encoded by *rpoC1* and *rpoC2*, respectively) (6).

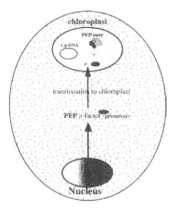

Fig. 1. Chimeric Composition of PEP holoenzyme Cp-DNA is a circular, double-stranded chloroplast (plastid) DNA. Precursor proteins encoded by the nuclear-genes are translated in the cytoplasm and then translocated into chloroplast. PEP sigma factor precursors are translocated into plastid to make up the functional PEP holo-enzyme.

The core subunit genes identified in the plastid genome show the β' subunit is split into two polypeptides, named β' and β'', exactly corresponding to the β' and γ subunits of cyanobacterial RNA polymerase. This observation is further support to the factuality that cyanobacteria and plant plastids are on the same lineage of the phylogenic pedigree. In spite of extensive physiological and biochemical studies of chloroplasts, we are not able to depict the overall profile of intergenomic interplays between plastid genome and nucleus mainly because of the absence of genetic background to the contribution of nuclear genes for the function and differentiation of plastids.

Recently it has been proposed that at least two types of transcription machineries, plastid-encoded RNA polymerase (PEP) and nuclear-encoded RNA polymerase (NEP) are involved in the transcription in plastid genome (7, 8). PEP is a prokaryotic-type RNA polymerase that consists of two portions, the multi-subunit core enzyme and the sigma factor, comparable to the eubacterial transcription machinery (Fig. 1). All of the genes encoding for the core subunits of PEP enzyme are found in the plastid genome, but a pivotal protein of eubacterial-type RNA polymerase, σ factor, is not found. Plastids originated by an endosymbiotic event still maintain a number of prokaryotic features, although they are now an integral part of plant cells. In any event, so as to understand the coordinated regulatory mechanisms between plastids and nucleus, it is crucial to learn a lot of things from cornucopia of eubacterial molecular biology.

2. PRINCIPAL-TYPE SIGMA FACTORS IN CYANOBACTERIA AND PLASTIDS

In this section we describe biological significance of principal-type sigma factors in cyanobacteria and successful isolation of nuclear genes encoding plastid sigma factors from primitive red algae, *Cyanidium caldarium* RK-1.

2.1. Eubacterial RNA polymerase sigma factors

Transcription machinery of cyanobacteria is a slight variant of canonical eubacterial RNA polymerases consisting of the multi-subunit core enzyme (E) and a σ factor. In *E. coli*, it is known that seven different sigma factors, each of which has its own promoter-specificity, work as a modulator of transcription by substituting with the principal sigma factor, σ^{70}. The promoter selectivity of RNA polymerase holoenzyme (Eσ), positively controlled by replacing σ factors, is the primary mechanism prevailing in eubacterial gene expression controls. For example, RNA polymerase σ factor (σ^{70}) is replaced by σ^{38}, the *rpoS* gene product at stationary phase (9). In contrast to other minor σ factors, Eσ^{38} holoenzyme recognizes a number of *E. coli* promoters that are also recognized by Eσ^{70} (10). *E. coli* promoters can be classified into three groups: those recognized only by RNA polymerase holoenzyme containing σ^{70} (Eσ^{70}); those preferentially recognized by that containing σ^{38} (Eσ^{38}); and those recognized both Eσ^{70} and Eσ^{38}.

σ^{38} shares a common structure with the principal σ factors of divergent eubacteria and is a member of the RpoD-related protein family (11,12). The *rpoD box* sequence that is highly conserved between the principal sigma factors of *E. coli* and *B. subtilis* was used to design a versatile DNA probe to screen genes encoding for the principal or principal-type sigma factors from divergent eubacterial strains (11) (Fig. 2). As a result of survey of genes for the principal- and principal-type sigma factors, multiple sigma factor genes were identified in most of eubacterial strains in divergent lineages, including Gram-positive *Streptomyces* strains and Gram-negative cyanobacterial strains. In *Anabaena* sp. strain PCC7120, a gene (*sigA*) encoding the principal σ subunit, was cloned and analyzed (13). Two other genes (*sigB* and *sigC*) encoding closely related sigma factors were also identified in the same strain (14). Meantime, several other cyanobacterial strains, such as *Synechococcus* sp. strain PCC7942 (15), *Synechococcus* sp. strain PCC7002 (16), and *Microcystis* (*Synechocystis*) *aeruginosa* K-81 (17), were found to possess multiple genes encoding principal-type sigmas. More recently, the genome analysis of *Synechocystis* sp. strain PCC6803 identified several σ subunit genes on its genome (18).

On the basis of multiple sigma factors in eubacteria, Lonetto et al. (12) proposed to categorize the principal-type sigma factors as 'group 2'sigmas, which have essentially the same domain composition as those of the principal sigma factors (group 1) from different eubacteria strains. Here we refer the group 2 sigma factors as the principal-type sigma factors.

Although presence of principal-type sigma factors is common among eubacterial strains in divergent lineages, there are only a few lines of evidence supporting that these sigma factors (group 2 or principal-type sigma factors) have specific function(s) in the expression or regulation of specific class of genes. Group 1 as well as group 2 sigma factors of cyanobacteria possibly direct transcription initiation from the eubacterial consensus-type promoters containing the –10 element. Specificity crosstalk appears to be a common feature among eubacteiral group 1 and group 2 sigma factors (19). The stationary-phase or osmotic-pressure specific sigma factor, the *rpoS* gene product (σ^{38}) of *E. coli* confers the core enzyme to recognize specific promoters including the canonical consensus promoter sequence by *in vitro* transcription experiments (10).

2.2. Nuclear-encoded chloroplast RNA polymerase sigma factor in a red algae
Cyanidium caldarium RK-1

In the previous section, we discussed that extant plants have evolutionarily acquired the ability to perform the oxygen-generating photosynthesis by endosymbiosis of cyanobacteria, although the endosymbiotic events predated far before the establishment of land plants lineages. Nevertheless, we are able to realize the vestiges of the ancient endosymbiotic origin of plastids by the remarkable similarities between functions, structures and genetic constituents of plastid genomes and those of cyanobacteria. The key point to demystify the coordination of gene expression between nuclear and plastidic genomes is to find the nuclear gene(s) encoding the sigma factor of the PEP RNA polymerase that is pivotal in the transcriptional regulation.

Attempts to identify the sigma factor(s) functioning with the RNA polymerase core enzyme in plant plastids by biochemical or immunological approaches have been frustrated by the fact that proteins observed failed to reach definitive evidence at the gene level. Our approach to identify the gene(s) encoding the putative sigma factor that functions in connection with the core RNA polymerase of plastids was based on the two lines of evidence. The first is due to the fact that the genomes of cyanobacteria are highly similar to those of plant plastids. The second depends on the fact that unicellular red algae is apparently most primitive among extant plant cell lineages, and therefore, the nuclear gene(s) for the plastid sigma-like factor in query retains the similarity to the ancient cyanobacteria. Naturally, sequence information of cyanobacterial sigma factors is expected to serve an ideal source for the quest.

A strategy to amplify a specific nuclear region by a PCR was adopted to identify a gene encoding the RNA polymerase σ factor of the *Cyanidium* chloroplast. DNA primers which had been designed based on amino acid sequences conserved among the principal σ factors of cyanobacteria were synthesized and used to amplify the possible chloroplast sigma factor-encoding region of nuclear DNA (20). A single DNA region that hybridized with the probe was cloned in *E. coli*. The nucleotide sequence of the 4508-bp *EcoR-Bam*HI region revealed an open reading frame that encodes a putative protein homologous to cyanobacterial principal σ factors. Accordingly, we named this open reading frame *sigA*.

A. **Basic structure of eubacterial principal sigma factors**

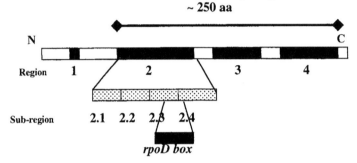

B. ***rpoD box* sequences of eubacterial and plastidic sigma factors**

Eubacterial Principal Sigma Factors	KFSTYATWWIRQA
Red Algae SigA (*Cyanidium caldarium* RK-1)	RFSTYATWWIRQS
Higher Plant SigA (*A. thaliana*)	KFSTYAHWWIRQA

Fig. 2. Schematic representation of functional regions in eubacterial principal sigma factor (A) and *rpoD box* sequences of eubacterial and plastidic sigma factors (B). Regions 2~4 spanning about 250 amino acid (aa) residues are functional domains proposed by Helmann and Chamberlin (5).

Localization of the *sigA* gene was confirmed by preferential hybridization with nuclear DNA. On the basis of high GC content of the *sigA* gene (50.4%) compared with the extremely low GC content of mitochondrial DNA, it is unlikely that the *sigA* gene is encoded by mitochondria. Thus, *sigA* appears to be encoded by the nuclear genome.

Several generally conserved domain structures, from NH2-terminal region 1.1 to COOH-terminal region 4.2, have been identified in σ proteins. The putative amino acid sequences of *sigA* gene product could be aligned from region 1.2 to the COOH-terminus with the sequences of principal σ factors of cyanobacteria. Sequence similarity continues in a portion between regions 1.2 and 2, which is fairly less similar among eubacterial σ factors, indicating that the putative SigA protein appears to be a close relative of cyanobacterial principal sigma factors (Fig. 2). In general, nuclear-encoded chloroplast proteins are synthesized as larger precursors containing an NH2-terminal extension called a transit peptide (21). A serine- and threonine-rich sequence, which possibly functions as a transit peptide, was found in the NH2-terminal region.

To determine whether SigA can function as a eubacterial σ factor, we purified the recombinant protein after expression in *E. coli*. The expressed protein contains a leader sequence and His-tag derived from the expression plasmid. The purified protein was reconstituted with core RNA polymerase of *E. coli* and the transcriptional specificity of the resulting enzyme was examined *in vitro*. The heterologous holoenzyme recognized the consensus *E. coli* promoters for *tac* and *RNA I*, confirming that SigA can function as a eubacterial principal or principal-type σ factor.

3. IDENTIFICATION OF GENES FOR PLASTID RNA POLYMERASE SIGMA FACTORS FROM HIGHER PLANTS

As described in the previous section, we succeeded in demonstrating that a plastid sigma subunit is actually encoded by the nuclear genome in primitive red algae *Cyanidium caldarium* RK-1 (18). Therefore, it is highly likely that genes in higher plant nuclei also encode for precursors of plastid sigma factors, which are targeted into plastid to control the plastid gene expression. In this section, we describe the basic characterization of cDNA species encoding plastid sigma factors from three higher plants, two dicotyledonous, *Arabidopsis thaliana* and *Nicotiana tabacum*, and one monocotyledonous, *Oryza sativa*.

3.1. *Arabidopsis thaliana*

We identified nuclear genes encoding plastid RNA polymerase sigma factors by initial database search analyses and corresponding cDNA cloning (20). First, three *Arabidopsis* EST (Expression Sequence Tag) entries having similarity to eubacterial RNA polymerase sigma factors. cDNA clones corresponding to these partial sequences were isolated, and the complete nucleotide sequences were determined. All three sequences encode proteins highly similar to plastid sigma factors in cyanobacteria and red algae, and the gene products have N-terminal extensions, which presumably function as plastid-targeting transit peptides. Thus we concluded that the gene products are RNA polymerase sigma factors of plastids, and named *sigA*, *sigB* and *sigC*, respectively. Expression of these genes was analyzed by RNA gel-blot analysis and induced by light illumination after a short-term dark adaptation. This strongly suggests that light regulation of the nuclear encoded sigma factor genes would be involved in light-dependent activation of plastid promoters. This is the first isolation of cDNA species for RNA polymerase sigma factors of higher plant plastids.

112

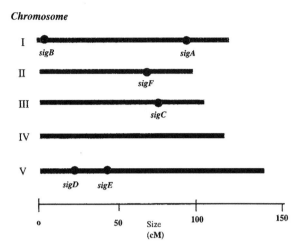

Fig.3. Approximate locations of six *sig* genes on *A. thaliana* chromosomes.

More recently, another three EST entries have been confirmed to encode plastid sigma factor-like protein (23). Cloning and sequence determination of corresponding cDNAs indicated three new *sig* genes, designated *sigD*, *sigE* and *sigF*. These three genes were estimated to encode polypeptides of 419, 517 (or 543) and 553 amino acids, corresponding to molecular sizes of 47.2, 58.8(or 61.6) and 62.5 kDa, respectively. Chromosomal locations of the six *Arabidopsis sig* genes are schematically shown in Fig 3. We note here that *sigD* and *sigE* are identical to the gene *SIG4* (AF101075) and *sig5* (Y18550), discussed by other (24).

These six *sig* gene products deduced from the cDNA sequences belong to group 1 or group 2 sigma factors of eubacteria. The structures of the promoter recognition domains, region 2.4 and 4.2, highly resemble each other, indicating that these sigma factors recognize similar promoter sequences. In *E. coli*, group 1 σ^{70} and group 2 σ^{38} proteins recognize similar promoter structures sharing the consensus 'TATAAT' type –10 element. Since most plastid promoters also contain a similar promoter structure, involvement of group 1 or group 2 sigma proteins has been suggested. Thus, it coincides well with the presence of the highly homologous sigma genes in higher plants. These sequence data do not provide any direct evidence that the gene products function as sigma factors of plastid. However, the sequence similarity to cyanobacterial and plastid sigma factors, as well as the presence of N-terminal plastid targeting sequences, strongly suggests that all of them are nuclear encoded plastid sigma factors.

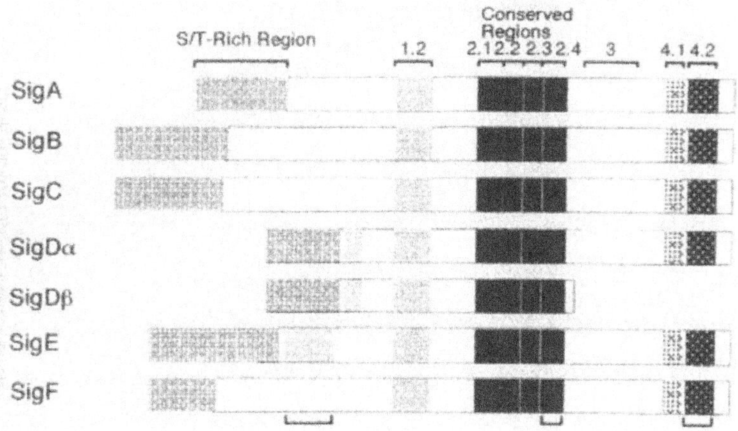

Fig.4. Schematic representation of six sig gene products of *Arabidopsis thaliana*.

Two alternative spliced products of *sigD* gene (SigD α and SigD β) are also shown. S/T-rich regions are possible N-terminal plastid-targeting peptides. Conserved regions corresponding to the functional domains of eubacterial principal sigma proteins are also shown.

Transcript analyses revealed two alternatively spliced transcripts generated from the *sigD* region, one of which is predicted to encode a σ protein lacking the carboxy-terminal region 3 and 4. Finally, the amino-terminal sequence of the *sigF* gene product was shown to function as a plastid-targeting signal using green fluorescent protein fusions (Fig. 4).

Transcripts of the three-*sig* (*sigA*, *sigB* and *sigC*) genes were similarly induced by light illumination after 24-h dark adaptation. Some promoters are induced by light in plastids, and among them, the *p3* promoter of the *psbS/C* operon is best characterized. Expression of *sig* genes responded to developmental stages of seedling would be analysed and discussed in the other section.

Phylogenetic analysis revealed that many of higher plant σ factors fell into at least four distinct subgroups within a diverse protein family. Each subgroup corresponds to *A. thaliana* SigA, SigB, SigC, and SigF. Furthermore, SigA subgroup members were further divided into monocot and dicot groups (23). In addition, *Arabidopsis sig* genes contained conserved chromosomal intron sites, indicating that these genes arose by DNA duplication events during plant evolution.

3.2. Tobacco and Rice

Tobacco (*Nicotiana tabacum*) is a typical dicotyledonous plant that has been used especially for plastid researches as a versatile model plant. The first complete nucleotide sequence of plastid genome in higher plants was determined in tobacco (2,3). Since then, a number of plastid-encoded genes were characterized at the molecular level. Furthermore, cultured cell lines and the stable transformation system of plastid genome in tobacco have been established and widely available. Therefore, characterization of tobacco plastid σ factors may facilitate the understanding of the regulation of plastid gene expression. We have isolated and characterized two genes (*sigA1* and *sigA2*) from *N. tabacum*, whose products presumably function as σ factors for plastid RNA polymerase (25). Transient expression assay using a GFP (Green Fluorescent protein) fusion construct indicated that the N-terminal region of the *sigA2* gene product could function as a transit peptide for import into chloroplasts. The gel-blot analysis of RNAs revealed that the sum of the *sigA1* and *sigA2* transcripts fluctuated apparently with an endogenous rhythm after 12-hr-light, 12-hr-dark entrainment in photmixotrophically cultured tobacco cells.

Screening of an *Oryza sativa* cDNA library with a probe derived from the PCR product of *A. thaliana* cDNA library was successfully used to identify cDNA clones corresponding to the *O. sativa* sigma factor (26). Among eight such clones, the longest one that gave the complete 2038-bp nucleotide sequence was named *OssigA*. (*Oryza sativa* sigma factor A). The *OssigA* encodes a 519 amino acid polypeptide, whose amino acid sequence was 50% identical to that of *A. thaliana* SigA. The N-terminal region of OsSigA, showing especially high homology to AtSigA, follows by a stretch of 13 glycine residues and a putative cleavage site (AVAA) that shows similarity to that of the plastid-targeting sequence LTVVAA. The predicted amino acid sequence of OsSigA showed substantial homology to conserved regions 1.2-4.2 and is highly conserved among microbial principal σ factors, was also especially well conserved in the σ factor-like polypeptides of higher plants and *C. caldarium* RK-1.

Southern blot analysis of genomic DNA from *O. sativa* revealed that *OssigA* is a single-copy nuclear gene. A probe corresponding to the cDNA sequence encompasses regions 1.2-42. Of the encoded protein yielded specific signals. Expression of *OssigA* was markedly greater in the shoot than in the root of rice seedlings grown under a normal light/dark cycle. Furthermore, whereas the abundance of *OssigA* transcripts was markedly reduced in the shoot of etiolated seedlings grown in the dark, it increased rapidly on exposure of such shoots to light. Thus, the *OssigA* expression appears to coincide with that of light-dependent chloroplast biogenesis. Light induces marked changes in plant cells that are accompanied by coordinated transcription of genes in both the nucleus and plastids.

4. FUNCTIONAL ANALYSES OF *SigA* AND *SigB* GENES IN *Arabidopsis thaliana*

To date nucleotide sequences encoding higher plant sigma factors have been identified from various higher plant cells, such as rice (26), mustard (27), maize (28), wheat (29), and tobacco (25). However we do not have direct evidence that these nuclear-encoded sigma-like proteins are functioning as a transcriptional initiation factor in plastids. The simplest and straightforward explanation of these findings is that the eubacterial sigma factor-like protein function as a transcriptional initiation factor with the PEP core enzyme in plastids. Eubacterial σ factors are categorized as a positive control factors which can modulate expression of a number of genes altogether. Thus, if the nuclear-encoded plastid sigma factors function in plastids as expected from the structures, presumably the nucleus is capable of governing PEP-dependent plastid gene expression in the coordination with cell growth, metabolism, and development, mediated by control of the widespread sigma factors. Here we outline the genetic and physiological characteristics of *Arabidopsis sig* genes (30,31).

4.1. *sigA* and *sigB* expression in green tissues

We have identified six nuclear genes encoding for putative PEP sigma proteins from *Arabidopsis thaliana* (*sigA ~ sigF*). Preliminary northern blot analysis has indicated that *sigA* and *sigB* genes are more abundantly expressed under the normal growth condition. It appears that the *sigA* and *sigB* gene products possess the most important roles for PEP-dependent transcription in the *Arabidopsis* chloroplasts.

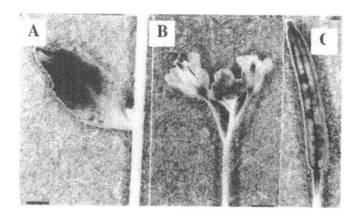

Fig. 5. Tissue-specific activation of *sigB*-promoter in *A.thaliana* plants. Panel (A), cauline leave; panel (B), flowers; panel (C), silique.

To know the expression patterns, the 5' upstream regulatory regions from *sigA* and *sigB* genes were fused with the *uidA* gene on Ti-plasmid pB110, and the transgenic *A. thaliana* lines were constructed. Both the resultant lines, named AelG and BelG, were shown to express a transcriptional fusion which encodes β-glucronidase (GUS) with no additional amino acids at the N-termini. The homozygous transgenic lines were established and cultivated under a continuous light condition to examine the tissue specific expression from *sig* genes. As the result, rosette and cauline leaves, seals, and siliques were stained, but seeds or other flower-organs were not with X-Glux. Mature rosette sepals and cauline leaves were stained entirely but obscurely. Apparently, the staining patterns were at random and not related to the age or position of the leaves. The basement cells of trichomes were most densely stained and the trichome themselves were not generally stained. Although mature siliques were entirely stained, smaller or immature siliques were only stained partly at the top. In contrast, stems beneath flowers or siliques were particularly well stained (Fig. 5).

4.2. *sigA*- and *sigB*-promoter activities at cotyledons and the first rosette leaves

The next major purpose of our experiment was to trace the early activation of the *sigA*- or *sigB*-promoter during seed germination after release from cold treatment. When the imbibed seeds of the transgenic lines were placed under a dark condition and then germinated under continuous light, most of the seed coats were torn and then about 24 to 36 h after the release from the cold treatment, and the cotyledons fully opened by the third day. After these events, vigorous expansion of the cotyledons for a week was observed relating to the development of chloroplasts. The first rosette leaves had mostly appeared by the sixth day.

When the whole seedlings of *sig-* and *cab3-uidA* fusion were strained every 24-h genesis, no detectable GUS activity was found in imbibed seeds just after the release. Unlike the mosaic (non-uniform) staining of rosette leaves or cauline leaves, all cotyledons were stained uniformly (Fig. 6). The first detectable activity of *sigA*-promoter was recognized at least one day later from that of *sigB*-promoter activity. In the seven-day-old plants, both *sigA-* and *sigB*-promoters were apparently activated in the first pair of rosette leaves from the marginal area inward. This result indicates that the *sigB* gene is expressed earlier at least 24 h before *sigA* gene.

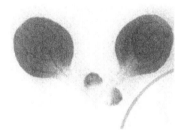

Fig. 6.   Uniformly activated *sig*-promoter in cotyledons of *A. thaliana*.

4.3.   Transient and repeated activation of the *sigA*- and *sigB*-promoters in young seedlings

To quantify the promoter activities in whole young seedlings during chloroplast biogenesis, we traced the kinetics of GUS activities in *sigA-* and *sigB-uidA* transgenic strains with every 12-h for five days after the release. The *sigB*-promoter activity was detected prior to the *sigA*-promoter activity, which is consistent with the previous results (Fig. 7). The first peaks of the transgenic *Arabidopsis* lines were similarly achieved at around 48 h after the release from cold treatment and remarkably diminished within next 12 to 24 h. The second peaks of the two promoter activities emerged after 84 hr in the same way. Intriguingly, the second peak was drastically reduced to less than 10% within the next 12 h. A similar amplitude was observed during the next 24 h (108 and 120 h), as if the promoter activities oscillated with a 24 h cycle.

In the GUS assays of the transgenic *Arabidopsis* lines introducing *sigA*- or *sigB*-promoter *uidA* fusion, the *sigA*- and *sigB*-promoters were mainly activated in green tissues in adult plants, while those were entirely activated in young seedlings (Fig.7). These results suggest that the sigma factors are necessary for the early developmental stages of chloroplasts in cotyledons and young leaves. Furthermore, the differences of *sigA*- and *sigB*-promoter activities in adult leaves and young seedlings suggest that these genes are activated depending on not only the chloroplast development in cotyledons and young leaves but also the maintenance of photosynthetic functions in mature chloroplasts. Naturally chloroplast development and the maintenance of photosynthetic functions occur corresponding to various internal or external signals such as hormones or light.

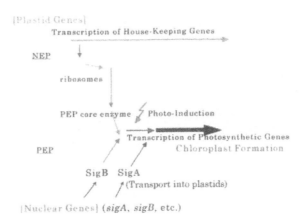

Fig.7.   Sequential activation of *sigB*- and *sigA*-promoters during chloroplast development in young seedlings.

4.4. The nuclear-encoded transcription factor SigB is required for the chloroplast development in *Arabidopsis thaliana*. (31)

It is known that light, presumably blue light induces the development of plastids into chloroplasts, organelles of specialized for photosynthesis. At moment, we do not know the factors involved in the light-induced plastid gene expression. There should be a complex cooperation among the relevant genetic events controlled by both nuclear and plastid genomes. As described in previous sections, at least six nuclear genes have been identified to encode plastid sigma factors, all of which were shown to possess possible plastid-targeting peptides in the N-terminal region. Two of such genes, *sigA* and *sigB*, were activated by a similar fashion at various plant tissues, such as cotyledons, hypocotyls, rosette and cauline leaves, sepals and siliques, but not at roots, seeds, or other flower organs. Furthermore, both the *sigA* and *sigB* promoters were repeatedly activated in young seedlings under continuous light, possibly in an oscillated fashion. Thus, it is likely that the nuclear gene-encoded sigma factors, SigA, and SigB of *A. thaliana*, have a critical role especially at an early stage of seedling growth in chloroplast development. Also, other sigma factors appear to contribute to

118

this process in a somewhat redundant manner.

The SigB protein and other sigma factors function as a link between the nuclear genome and plastid gene expression in higher plants. In a T-DNA insertion screen for mutations in chloroplast development an *A. thaliana* mutant, designated *abc1*, was shown to exhibit aberrant chloroplast development and a consequent phenotype characterized by pale green leaves. This phenotype was also biochemically quantified, revealing that the 4-day-old *abc1* cotyledons contain only 15% of the chlorophyll a/b content of 4-day-old wild-type cotyledons. Intriguingly, although the growth of this mutant is impaired, the macroscopic morphology of this mutant appears normal in all other aspects. Further characterization of this mutant by electron microscopy revealed that the developing chloroplasts in 5-day-old cotyledons of mutant plants are much smaller than the corresponding wild-type cotyledons. The thylakoid membranes and the stacked lamellar structure of grana were poorly developed, and few starch granules were apparent, suggestive of a low photosynthetic ability. In contrast, the chloroplasts from 5-day-old cotyledons of wild-type plants had substantial deposition of starch granules and the lamellar structure was far more developed. The size and shape of etioplasts in cotyledons grown in the dark were indistinguishable between mutant and wild-type plants. These observations thus indicated that the phenotype of the *abc1* plants results from a mutation that affects light-induced development of chloroplasts in the early stage of seedling growth (Fig.8).

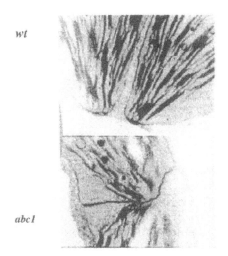

Fig. 8. Aberrant chloroplast development in *sigB*-disrupted (*abc1*) mutant of *A. thaliana*. Confocal laser-scanning microscopic observation of chlorophyll fluorescence in mesophyll cells of 2-week-old leaves of wild-type (left, *wt*) and *abc1* mutant (right). (Photo by Dr. Shimada of Tokyo Institute of Technology).

5. CONCLUSION

By scrutinizing the epigraphs depicted in the common sequences of the regulatory factors in the primeval inventor of oxygen-generating photosystems, cyanobacteria, we have successfully identified nuclear genes (*sig*) encoding for plastid sigma factors, first from unicellular red algae, *Cyanidium caldarium* RK-1. The same strategy was adopted to identify corresponding *sig* genes from three higher plants, two of typical dicotyledonous, *Arabidopsis thaliana* and *Nicotiana tabacum*, and one of monocotyledonous, *Orysa sativa*. Open reading frames found in the cDNA clones of these nuclear genes indicate that the N-terminal regions of the gene products had amino acid sequences typical to the plastid-targeting transit peptides. Furthermore, a transient expression assay of GFP fusions in *Arabidopsis* protoplasts showed that the N-termini of all three *sig* gene products functioned as chloroplast-targeting signals. The *sigA*- or *sigB*-promoter fused with *uidA* reporter in the transgenic *Arabidopsis*, were similarly activated at various tissues of the plants, such as cotyledons, hypocotyls, rosette leaves, sepals and siliques, but not at roots, seed, or other flower organs. Promoters including those from *Cyanidium*, *Arabidopsis*, and *Nicotiana* were repeatedly activated under continuous light, somewhat similar to endogenous rhythms. An *Arabidopsis* mutant (*abc1*) having a pale-green leaf phenotype presumably by the impaired *sigB* function was isolated as a T-DNA insertion clone. This result provides direct evidence that a nuclear-derived prokaryotic-like SigB protein plays a critical role in the coordination of the two genomes for chloroplast and plastid development.

REFERENCES

1. M. Avron, Curr. Top. Bioenerg. 2(1967) 1.
2. K. Shinozaki et al., EMBO J., 5(1986) 2046.
3. M. Sugiura, Plant Mol.Biol., 19 (1992)149.
4. J. Bauer et al., Nature 403(2000) 203.
5. J. D. Helmann and M.J. Chamberlin, Annu. Rev. Biochem. 57(1988) 839.
6. G.L. Igloi   and H. Kossel (1992) Crit. Rev. Plant Sci. 10: 525.
7. P.T.J. Hajdukiewicz, L.A. Allison and P. Maliga, EMBO J. 16 (1997) 4041.
8. P. Maliga, Trends Plant Sci., 3 (1998) 4 .
9. R. Lange and R. Hengge-Aronis, Mol. Microbiol., 5 (1991) 49.
10. K. Tanaka, Y. Takayanagi, N. Fujita, A. Ishihama and H.Takahashi, Proc. Natl. Acad. Sci. USA, 90 (1993) 3511.
11. K. Tanaka, T. Shiina and H. Takahashi, Science 242(1988) 1040.
12. M.Lonetto, M.Gribskov and C. Gross, J. Bacteriol., 174 (1992) 3843.
13. B. Brahamsha and R. Haselkorn, J. Bacteriol., 173 (1991) 2442.
14. B. Brahamsha and R. Haselkorn, J. Bacteriol., 174 (1992) 7273
15. K. Tanaka, S. Masuda and H. Takahashi, Biochim. Biophys. Acta, 1132 (1992) 1132.
16. L.F. Caslake and D.A. Bryant, Microbiol., 142 (1996) 347.
17. M. Asayama, A. Suzuki, S. Nozawa, A.Yamada, T. Aida, K. Tanaka, H. Takahashi and M. Shirai, J. Biochem. 120 (1996) 752.
18. T. Kaneko, S. Sato, H. Kotani, K. Tanaka, E. Asayama, T. Kamura et. al., DNA Res. 3: (1999) 109.
19. A. Goto-Seki, M. Shirokane, S. Masuda, K. Tanaka and H. Takahashi, Mol. Microbiol., 34(1999) 473.

20. K. Tanaka, K. Oikawa, N. Ohta, H. Kuroiwa, T. Kuroiwa and H. Takahashi, H., Science 272 (1996) 1932.

21. K. Keegstra, L.J. Olsen and S.M. Theg, Annu. Rev. Plant Physiol. 40(1989) 471

22. K. Tanaka, Y. Tozawa, N. Mochizuki, K. Shinozaki, A. Nagatani, K. Wakasa and H. Takahashi, FEBS Lett. 413 (1997) 309.

23. M. Fujiwara, A. Nagashima, K. Kanamaru, K. Tanaka and H. Takahashi, FEBS Lett. 141(2000) 47.

24. M.A. Hakimi, I. Privat, J.G. Valay and S. Lerbs-Mache, J. Biol. Chem., 275 (2000) 9215.

25. K. Oikawa, M. Fujiwara, M. Nakazato, K. Tanaka, K. and H. Takahashi, Gene 261(2000) 221-228.

26. Y. Tozawa, K. Tanaka, H. Takahashi, H. and K. Wakasa, Nucl. Acids Res., 15(1997) 415.

27. M. Kestermann, S. Neukirchen, K. Kloppstech and G. Link, Nucl. Acids Res. 26 (1998) 2747.

28. S. Tan and R. F. Troxler, Proc. Natl. Acad. Sci. USA 96 (1999) 5316.

29. K. Morikawa S. Ito, Y. Tsunoyama, Y., T. Shiina T and Y. Toyoshima. FEBS Lett. 451 (1999) 275.

30 K. Kanamaru F. Fujiwara. M., Seki M., Katagiri T., Nakamura M., Mochizuki, N., A.Nagatani, K. Shinozaki, K. Tanaka and H. Takahashi, Plant Cell Physiol., 40 (1999) 832.

31. Y. Shirano, H. Shimada, K. Kanamaru, M, Fujiwara, K. Tanaka, H. Takahashi, K. Unno, S. Sato, S.Tabata, H. Hayashi, C. Miyake, A. Yokota and D. Shibata. FEBS Lett. 485 (2000) 178.

Molecular Anatomy of Cellular Systems
I. Endo et al., (editors)

From viral RNA genome to infectious ribonucleoprotein complexes through RNA replication

Kyosuke Nagata*

Department of Molecular Medical Engineering, Graduate School of Bioscience and
Biotechnology, Tokyo Institute of Technology
4259 Nagatsuta, Midori-ku, Yokohama 226-8501, Japan

A virus virion is a chemical complex consisting of either DNA or RNA as its genome, proteins, and a membrane fraction derived from host cells. A virion itself never replicates in a test tube but does in cells. Thus, a virus is not defined as an organism but rather a parasitic substance that can be amplified depending on host cell functions. In infected cells, transcription and replication of the virus genome take place, and amplified genomes are assembled into progeny virions. To know molecular mechanisms of these processes, not only forward- and reverse-genetics but also biochemical approaches have been thought needed. Here we describe the summary of our recent efforts to clarify the mechanism of replication and transcription of a viral genome and of formation of viral genome-protein complexes prerequisite for progeny virion formation. Using influenza virus, an RNA virus, as a model virus, we first established *cell-free* systems in which viral RNA synthesis occurs, and then carried out dissection and reconstitution of the systems. It was shown that the viral RNA-dependent RNA polymerases and nucleoprotein are essential for viral RNA synthesis. We succeeded in reconstitution of functional viral RNA-nucleoprotein complexes. Interestingly, the RNA synthesizing activity was stimulated by host factors that could be categorized into 'acidic molecular chaperones'. It was indicated that progeny viral RNA-protein complexes are exported from nuclei through a leptomycin B-sensitive pathway to cytoplasm, where virion formation starts. Viral matrix protein is though to cover viral RNA-protein complexes and also function as templates for envelope formation with host cell-derived membrane fractions. In fact, the matrix protein was capable of interacting with envelope components as well as the viral RNA-protein complexes and viral RNAs. Roles and structure-function relationships of host cellular- and viral-factors involved in these processes are discussed.

1. INTRODUCTION

Influenza virion has eight segments of single-stranded RNA of negative polarity (vRNA)

*Present Address; Department of Infection Biology, Institute of Basic Medical Sciences, University of Tsukuba,
1-1-1 Tennodai, Tsukuba 305-8575,

122

as its genome, which are complexed with viral RNA-dependent RNA polymerase and nucleoprotein (NP) (Figure 1). A 12 nucleotide-long sequence and a 13 nucleotide-long sequence at 3' and 5' terminal regions of the genome RNA, respectively, are well conserved among genome segments. These 3' and 5' conserved regions are partially complementary and form the partial double stranded RNA. The partially hybridized terminal regions have been referred to as 'pan-handle' (1), 'fork', 'hook', and 'corkscrew' forms (2-5). Signals for transcription and replication of the influenza virus genome RNA are located in the conserved region (6). In the vRNA-protein complexes (vRNP complexes) prepared from purified virions, the viral RNA polymerases are found bound to the pan-handle region (6), and NP encoded by RNA5 is cooperatively bound to RNA segments every 15-20 nucleotides (7). The RNA polymerase is composed of three viral proteins, PB2, PB1, and PA, encoded by RNA1, RNA2, and RNA3, respectively. Transcription and replication of the viral genome take place in nucleus, to which incoming RNP complexes are transported after being dissociated from other viral components.

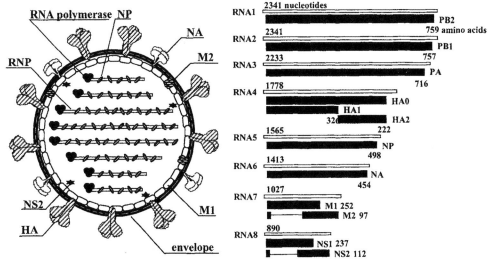

Figure 1. The structure of influenza virus virion and genome. The virion structure of influenza virus is schematically represented. The outline of the influenza virus A/PR/8/34 genome, RNA1-8, is shown (blank box) with number of nucleotides. The viral proteins encoded by RNA segments are indicated (black box) with number of amino acids.

RNP is surrounded with the membrane-associated matrix 1 (M1) protein encoded by RNA7. RNA7 encodes M2, which is generated from spliced RNA7. Two transmembrane glycoproteins, hemagglutinin (HA) and neuraminidase (NA) encoded by RNA4 and RNA6, respectively, exist through the envelope of virion consisting of host cell-derived membrane fractions. M1 is also likely to be associated with lipid bilayer (8). RNA8 codes for viral nonstructural protein (NS) 1 and NS2, the latter of which is translated from alternatively spliced RNA8 mRNA. NS1 and NS2 are proposed to function in the shutoff phenomenon of cellular translation and in nuclear export of progeny RNP complexes, respectively.

Exported RNP complexes from nuclei are assembled into virions on the cell membrane where HA, NA, and M2 are located.

Here we summarize our recent efforts to clarify the mechanism of replication and transcription of a viral genome, in particular with regard to factors derived from host cells (host factors) and that of formation of viral genome-protein complexes prerequisite for progeny virion formation.

2. INVOLVEMENT OF HOST CELL FACTORS IN INFLUENZA VIRUS RNA SYNTHESIS

Knowledge about the molecular mechanism for transcription and replication of the influenza viral genome is accumulating (6 and references therein), in particular about the function of viral proteins (Figure 2). One of the viral RNA polymerase subunits, PB2, recognizes a cap structure of cellular mRNA and generates an oligonucleotide containing the cap structure. Another subunit, PB1, is the catalytic subunit for viral RNA synthesis. The role of the other subunit, PA, is not known at present, although genetic experiments have suggested that PA is involved in the replication process. NP, an RNA binding protein, is required for efficient elongation of the RNA chain. Transcription, the process of producing viral mRNA, is initiated by recognition with PB2 of the cap structure of nuclear pre-mRNA. PB2 cleaves the capped RNA at 10-13 bases downstream from the 5' end. The capped oligoribonucleotide serves as a primer for viral mRNA synthesis catalyzed by PB1. Elongation of the RNA chain proceeds until the polymerase reaches a polyadenylation signal, consisting of 5-7 U residues located near the 5' terminal region of the vRNA. The viral RNA polymerase polyadenylates the nascent RNA chain possibly by a slippage mechanism at the U-stretch. Replication of the viral genome consists of two RNA synthesis steps. The first step is the synthesis of complementary RNA (cRNA) from vRNA, and the second step is the amplification of the progeny vRNA from cRNA.

The molecular mechanism of transcription and replication is becoming clear. Based on the information, infectious RNA/DNA systems (where "infectious RNA/DNA" means that the progeny virion is produced when RNA or cDNA corresponding to the viral genome with or without factors is introduced into cells) are established (9-11). However, this does not exclude a possibility that host factors are involved in transcription and replication. Indeed, cellular capped RNA, for instance, is the primer for transcription. We reported that there could be a factor(s) in nuclear extracts of influenza virus infected HeLa cells which is possibly involved in the switch of mRNA to cRNA synthesis (12). Furthermore, recent experiments using yeast two hybrid systems have identified cellular proteins that interact with NP and NS1 (6). To identify a host factor(s), we developed a *cell-free* system supporting transcription and replication with nuclear extracts prepared from influenza virus-infected HeLa cells using exogenously added model viral RNA templates (13). With this system in our hands, we decided to identify host factors, if any, by biochemically dissecting and reconstituting the system.

2.1. Identification of host factors that stimulate viral RNA synthesis

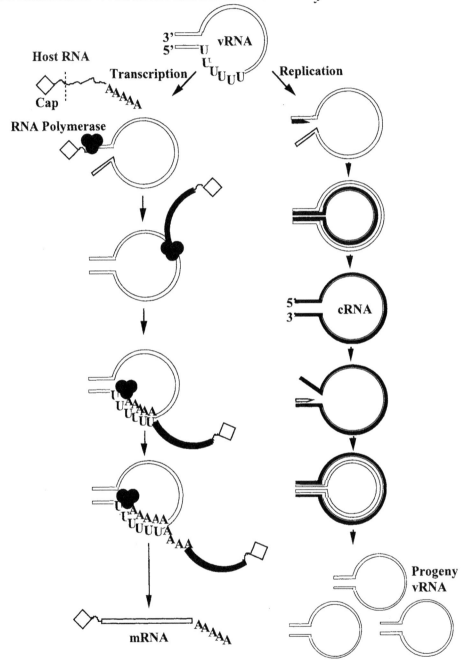

Figure 2. Transcription and replication of the influenza virus genome.

Nuclear extracts for the *cell-free* system was prepared from either influenza virus-infected or uninfected HeLa cells as described by Dignam *et al* (14). The 53 nucleotide-long model viral RNA contains terminal conserved sequences at the 3' and 5' ends of viral RNAs (13), so that it contains minimal signal sequences for transcription and replication.

Biochemical fractionation of infected HeLa nuclear extracts revealed that the maximal RNA synthesizing activity is reconstituted with two complementing fractions,the ribonucleoprotein (RNP) complexes and the fraction free of RNP (13). Of interest is that the former could be replaced with RNP complexes isolated from virions consisting of vRNA, the viral RNA polymerases, and NP, while the latter was substituted with nuclear extracts prepared from uninfected cells. These observations confirmed that virion-derived RNP complexes are the catalytic unit for RNA synthesis, and suggested that the RNA synthesizing activity is modulated by host-derived nuclear factors.

By fractionation of nuclear extracts from uninfected cells through phosphocellulose column chromatography (Figure 3), the stimulatory activity for RNA synthesis supported by RNP complexes was separated into two distinct fractions (13). The RAF (RNA polymerase activating factor) fraction that was not absorbed to a phosphocellulose column stimulated RNA synthesis with the RNA polymerase either associated with or free of vRNA. Using limited elongation assays, the stimulation by the RAF fraction was shown to operate at the initiation stage and/or the early elongation stage of RNA synthesis. The PRF (polymerase

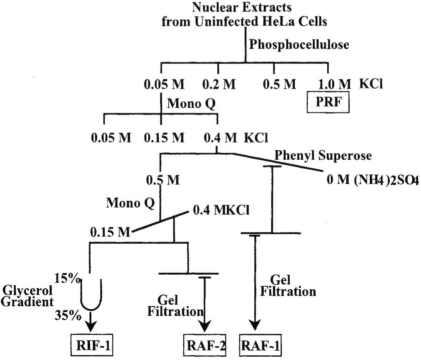

Figure 3. Purification scheme of host factors involved in influenza virus RNA synthesis.

regulating factor) fraction eluted from phosphocellulose by 1.0 M KCl stimulated RNA synthesis only when RNP complexes were used as the enzyme source.

The stimulatory activity present in the RAF fraction was subjected to further fractionation by column chromatography (15). The stimulatory activity was tightly bound to Q-sepharose. The stimulatory activity recovered from Q-sepharose was further separated through phenyl-sepharose column chromatography into two distinct fractions, unbound and bound fractions designated RAF-2 and RAF-1, respectively. Since RAF-1 and RAF-2 additively stimulated RNA synthesis, these activities were suggested to be distinct from each other. When these fractions were fractionated through a gel filtration column, RAF-1 and RAF-2 activities were recovered in fractions corresponding to the molecular mass of 350 kDa and 60 kDa, respectively. The RAF-1 fraction did not contain polypeptides of 350 kDa but smaller ones as judged by polyacrylamide gel electrophoresis in the presence of sodium dodecyl sulfate (SDS-PAGE), suggesting that RAF-1 may exist as oligomer in a physiological condition. Furthermore, the RAF-2 fraction was shown to contain an inhibitory activity, tentatively designated RIF-1 (RNA polymerase inhibitory factor-1). RIF-1 sedimented as fast as bovine serum albumin in glycerol density gradient centrifugation. RIF-1 was unlikely to be nuclease, since pre-labeled RNA remains unchanged during the *cell-free* RNA synthesis reaction.

2.2. Function of host factors

It is noted that RAFs could be highly negatively charged proteins due to their chromatographic behaviors. This protein nature led to the assumption that RAF functions as the stimulatory factor by interacting with a viral basic protein(s). To know the molecular identity of RAFs and their molecular roles and physiological functions in RNA synthesis of the influenza virus genome and in a cell, respectively, we decided to purify RAF proteins.

We first tried to purify RAF-2 (16), since we could hypothesize the action of RAF-2: RAF-2 gave less effect on RNA synthesis from pre-formed RNP templates than *de novo*-formed RNP templates with exogenously added RNA and RNA synthesis apparatuses, suggesting that RAF-2 interacts with the viral RNA polymerase and/or NP and facilitates their recruitment to naked RNA templates. The outline of the purification of RAF-2 is described in Figure 3. Analysis of the protein composition in the RAF-2 fraction of the final purification step by SDS-PAGE revealed that the RAF-2 fraction contained approximately equal amounts of a 48 kDa and a 36 kDa polypeptide, designated RAF-2p48 and RAF-2p36, respectively. RAF-2p48 and RAF-2p36 were individually eluted from the gel and subjected to renaturation. In the *cell-free* RNA synthesis system, the stimulatory activity was detected for RAF-2p48, but not RAF-2p36. This strongly suggested that RAF-2p48 is, at least, an active component of RAF-2.

To know the identity of RAF-2p48, oligopeptides prepared from RAF-2p48 were analyzed with matrix-assisted laser desorption-ionization time-of-flight mass spectrometry. The molecular mass data of the oligopeptides indicated that RAF-2p48 is identical with a known human protein, a 48 kDa putative ATP-dependent RNA helicase designated UAP56/BAT1. UAP56/BAT1 is found to interact with the linker region of splicing factor U2AF65 and appears to be an essential splicing factor required for the interaction of U2 snRNP with the pre-mRNA branchpoint (17). In fact, RAF-2p48 was localized in nuclei of uninfected cells, excluded from nucleoli and concentrated at spliceosomes where SC35, a splicing factor, is present. It was noted that the localization of RAF-2p48 is changed in influenza virus infected cells (see below). It is hypothesized that RAF-2p48/UAP56/BAT1 may be involved in conversion of RNA secondary structures within the U2 snRNA or the branch point (18), since it contains ATP-dependent DEAD-box RNA helicase motifs in its amino acid sequence. Independent of the purification of RAF-2, a yeast two-hybrid screen of a HeLa cell cDNA library using the influenza virus NP protein as bait identified RAF-2p48/BAT1/UAP56 as a NP interacting protein, designated NPI-5.

Next, we analyzed the function of RAF-2p48 in viral RNA synthesis using hexa-histidine- or GST-tagged recombinant RAF-2p48 proteins (His-p48 and GST-p48). We showed the His-p48 stimulates the cell-free viral RNA synthesis. In addition, a recombinant RAF-2p48 mutant protein containing a mutation in DEAD-box RNA helicase consensus motif VI, which is supposed essential for ATPase and helicase activities, showed the same level of activity as the wild-type His-p48. This result suggested that the putative RNA unwinding activity of RAF-2p48 might not be needed for stimulation of RNA synthesis by the recombinant RAF-2p48. To confirm the interaction between RAF-2p48 and NP, we carried out GST pull down assays using GST-p48 and either vRNP or micrococcal nuclease-treated vRNP (mnRNP), in the latter of which the RNA polymerases and NP are depleted of vRNA. GST-p48 specifically interacted with NP but not with the RNA polymerase when mnRNP was used. In contrast, GST-p48 did not interact with either NP or the RNA polymerase when vRNP was used. It was, therefore, concluded that RAF-2p48 binds to free NP but not to the NP-RNA complex. Furthermore, the addition of large amounts of RNA dissociated the RAF-2p48-NP complex, leading formation of RNA-NP complexes. Using mutant RAF-2p48 and NP proteins, domains responsible for their interaction were determined. RAF-2p48 was bound to the NP region consisting of amino-terminal 188 amino acid residues containing RNA recognition and binding domain. Further, NP interacted with RAF-2p48 through its carboxyl terminal region containing motif VI which is involved in RNA recognition. This result was supported by the yeast two-hybrid system with a series of NP deletion mutants. These observations suggested that formation of complexes either between RAF-2p48 and NP or between RNA and NP is mutually exclusive through competition for NP binding between RAF-2p48 and RNA. Finally, we examined the effect of RAF-2p48 on

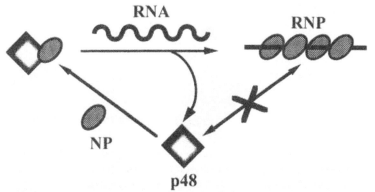

Figure 4. A proposed mechanism of the interaction among RNA, nucleoprotein, and RAF-2p48.

128

the efficiency of RNA-NP complex formation. Experiments using glycerol density gradients of mixture of NP and RNA in the presence or absence of RAF-2p48 revealed that formation of RNA-NP complexes is increased by RAF-2p48. Since the tertiary complex with RNA, NP, and RAF-2p48 was hardly detected, it was strongly suggested that RAF-2p48 functions as a chaperon for NP, thereby facilitating formation of NP-RNA complexes (Figure 4).

These findings brought us ideas that influenza virus smartly utilizes a host cell machine and that a chaperone may play an important role during formation of genome-protein complexes. With regard to the first one, RAF-2p48 was observed both diffusely and as concentrated speckles in nucleoplasm. A part of RAF-2p48 was shown co-localized with SC35, a splicing factor, so that it could exist in spliceosomes in uninfected cells. In contrast, in infected cells speckles of SC35 became small and separated. This was in good agreement with the observation that in infected cells, spliceosomes are partially destroyed. Of interest was that in infected cells, concentrated speckles of RAF-2p48 disappeared, and substantially less co-localization between RAF-2p48 and SC35 was seen. It has been suggested that in infected cells, splicing and 3' terminal processing of pre-mRNA are greatly inhibited (19, 20). The alteration of RAF-2p48 localization may be partially responsible for the splicing inhibition or concomitantly be caused by disruption of spliceosomes. Furthermore, re-localization of RAF-2p48 would contribute to the release of RAF-2p48 from spliceosomes and facilitate the recruitment of RAF-2p48 into the viral RNA synthesis machinery.

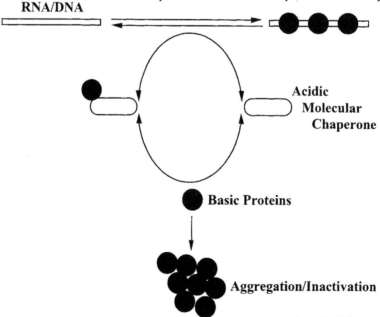

Figure 5. Acidic molecular chaperones. Formation of RNA/DNA-basic protein complexes and their dissociation are mediated by proteins containing the highly acidic region. These proteins may also prevent the aggregation/inactivation of basic proteins newly synthesized and/or transiently displaced from RNA/DNA.

In general, nucleic acid-binding basic proteins such as histones and viral basic proteins tend to aggregate and be inactive under physiological conditions in the absence of appropriate substrates such as DNA, RNA or possibly chaperones (6). Recently we have identified such the chaperone candidates in the course of experiments for reconstitution of a *cell-free* replication system of adenovirus chromatin (6). Proteins designated Template Activating Factors contains the highly acidic region which is essential for remodeling of the chromatin structure. We proposed the term 'acidic molecular chaperone' for these proteins (Figure 5). These proteins function as chaperone for the basic protein through their acidic property, so that they prevent the aggregation and inactivation of basic proteins and facilitate association and disruption of basic protein-nucleic acid complexes. On this line, it is interesting that a number of acidic cytoskeleton proteins is shown to be required for transcription and replication of the non-segmented negative strand RNA virus genome. Tubulin is a host factor for vesicular stomatitis virus (VSV), Sendai virus and measles virus, microtubule-associated proteins for VSV, and actin for measles virus and human parainfluenza virus type III (6). It is reported that the viral acidic P protein of non-segmented negative–sense RNA virus, which is required for viral RNA synthesis, forms complexes with NP and prevents inactivation of unassembled NP. Since influenza virus NP shows the protein nature similar to other viral NP, RAF-2p48 may participate in suppression of non-specific NP aggregation and may assist in delivery of NP to vRNA. In addition, RAF-1 is shown bound to one of viral basic proteins (unpublished observation). Although the identity of RAF-1 is presently not clear, it is quite possible that RAF-1 that contains a highly acidic region is also categorized into acidic molecular chaperone in influenza virus RNA synthesis.

3. RECONSTITUTION OF VIRAL RNA-PROTEIN COMPLEXES

To analyze the mechanism of virus multiplication including replication and transcription of the genome and to confirm findings obtained from *cell-free* studies, reverse-genetical methods are undoubtedly powerful. An RNA genome of negative polarity is incapable of generating infectious progeny virions upon introduction into cells. In the case of influenza virus, it was reported that transfection into cells of RNP complexes isolated from virions results in production of infectious progeny virions (21). Thus, reverse-genetics could be possible if RNP complexes including an engineered RNA are properly reconstituted. Furthermore, we are interested in the molecular mechanism of not only replication and transcription but also virion assembly. The process from genome replication to virion assembly is dissected into at least 5 distinct steps, *i.e.*, [1] for amplification of progeny RNAs, [2] for termination of RNA synthesis and formation of progeny RNP complexes, [3] for export of RNP complexes from nucleus to cytoplasm, [4] for coverage of RNP complexes with M1, and [5] for incorporation of RNP-M1 complexes inside envelope consisting host cell membrane fractions and viral spike glycoproteins. We have been trying to establish *cell-free* systems that are capable of reproducing the reaction of each step and then to anatomize and reconstitute the systems. By doing so, we could reconstitute infectious RNP complexes and RNP-M1 complexes. Furthermore, the mechanism of binding between M1 and membrane fractions (unpublished observations) and of nuclear export of RNP complexes has been studied (22).

3.1. Viral RNA-viral RNA dependent-RNA polymerases-nucleoprotein

The viral RNP complexes consist of vRNA, viral RNA-dependent RNA polymerases, and NP. One of important points for reconstitution of infectious RNP complexes is to construct a model viral RNA of negative polarity containing a marker gene so that the level of reconstitution is easily monitored. With such the model RNA, the strategy would be that gene expression and progeny virion production from the model RNA are examined by introduction of vRNA, vRNA-NP, vRNA-RNA polymerases, or vRNA-NP-RNA polymerases into cells with helper viral RNP complexes or helper virions (Figure 6).

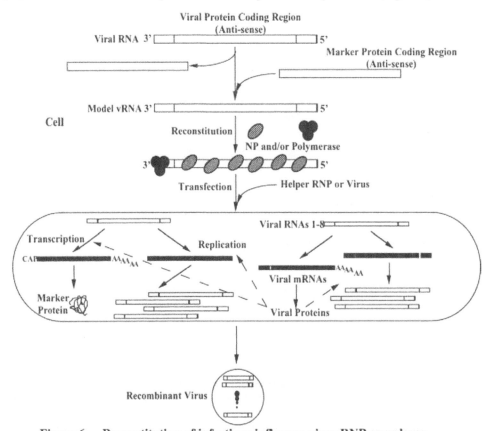

Figure 6. Reconstitution of infectious influenza virus RNP complexes.

We manipulated cDNA for RNA8 so as to replace the coding region of RNA8 with the *chloramphenicol acetyltransferase (CAT)* gene (10). The cDNA thus engineered was placed under the control of the promoter of T7 RNA polymerase. *Cell-free* transcription of *Mbo*II-digested plasmid by T7 RNA polymerase resulted in the production of the model viral RNA which contains, in order from 3' terminus to 5' terminus, 49 nucleotide-long virus-derived region, the antisense *CAT* gene ORF, and 35 nucleotide-long virus-derived region. Regions derived from the virus sequence contain signals involved in efficient translation in virus-infected cells (23).

We first tried to introduce into cells the naked model vRNA in the presence or absence of viral RNP complexes as helper by liposome-mediated transfection (10). Transcription from the model vRNA was examined by the indirect immunofluorescence assay with anti-CAT antibody and the RT-PCR methods. Transfection of naked vRNA resulted in no detectable level of CAT gene expression in both assays, confirming that RNA of negative polarity itself is not infectious. Further, when naked vRNA was introduced into influenza virus-infected cells, no significant amount of CAT was observed. This result indicated that naked model vRNA is incapable of functioning as templates for transcription. Since viral RNA polymerases and NP, essential *trans*-acting viral factors for transcription and replication, were to be synthesized in cells co-transfected with viral RNP complexes, the simple mixing of the model vRNA with viral RNP complexes could not be enough for maintaining the template activity of the model vRNA. It was possible that NP-vRNA, RNA polymerases-vRNA, or a complete RNP complex must be pre-formed prior to transfection with helpers.

NP was shown to be required for efficient RNA elongation (6), suggesting that RNA complexed with NP but not naked RNA may be a *bona-fide* template for the viral RNA polymerases. By characterizing the interaction between RNA and NP, we had succeeded in reconstitution of RNA-NP complexes structurally similar to native RNP complexes (7). For these reasons, we tested the activity of the model vRNA-NP complexes that had been reconstituted prior to transfection (10). By transfection of the pre-formed model vRNA-NP complex with helper RNP complexes, significant numbers of fluorescence-positive cells were detected. Total cellular RNA was prepared from transfected cells and subjected to RT-PCR using the model RNA-specific primers. No amplified DNA specific for positive-sense RNAs derived from the model vRNA was detected from RNA of cells transfected with helper RNP complexes alone. In contrast, PCR products were detected from model vRNA-NP complex- and helper complex-transfected cells depending on the presence of specific primers, reverse-transcription, and the cycle number of PCR. These observations indicated that transcription of the model vRNA takes place in transfected cells. Preliminary experiments revealed the amplification of the negative-sense model RNA and the production of progeny virions containing the model vRNA by this transfection method (unpublished observations). Luytjes et al. (9) reported the infectious RNP system with viral RNA polymerases and NP used for reconstitution. It is possible that NP is needed for a stringent template structure and RNA polymerase could be supplied in transfected cells. When adenovirus DNA binding protein or calf thymus histones, both of which can bind to RNA, is used instead of NP during reconstitution, only a limited extent of CAT expression was observed. This suggested that RNA-NP complexes form some particular and specific structure for recognition by the viral RNP polymerase.

It had been suggested that the transcription promoter resides in the 3' terminal region of vRNA that possibly forms the pan-handle structure with the 5' terminal region. We used this system for analyses of *cis*-acting regions involved in transcription. To this end, a series of mutants containing base substitutions in the 3' terminal region of the model vRNA was constructed and subjected to the transfection method. Mutations of the 3'-terminal base gave no effect on transcription, whereas mutations between nucleotide positions 6-14 from the 3' end significantly reduced the level of transcription. This suggested that signals important for transcription are located near the 3' end but not at the 3'-terminal base.

3.2. RNP-M1 complexes

The influenza virus matrix protein 1 (M1) has at least the dual binding property. M1 surrounds RNP complexes interacting with a component(s) of RNP complexes and is thought to function as a structural template for virion assembly with membrane fractions of infected cells that consist of viral transmembrane glycoproteins and lipid bilayer from the cell membrane. In this section, we describe the mechanism of RNA/RNP binding of M1 and M1-mediated inhibition of viral transcription (alternatively designated anti-RNA synthesis). The interaction mechanism between M1 and membrane fractions is discussed in the latter section.

M1 is dissociated from and assembled with RNP complexes, before incoming RNP complexes are transported into nucleus and when progeny RNP complexes are exported from nucleus to cytoplasm where progeny virion assembly initiates, respectively. Several lines of evidence show that M1 inhibits the cell-free RNA synthesis catalyzed by RNP complexes prepared from virions (24 and references therein). This concept is brought by cell-free experiments monitoring RNA synthesis from endogenous RNA templates associated to RNA complexes. We decided to study on the interaction mechanism between M1 and RNP complexes and to re-examine M1-mediated inhibition of transcription using our cell-free RNA synthesis system that is dependent on the exogenously added RNA template.

We prepared histidine-tagged recombinant M1 (His-M1) proteins including wild-type and a series of deletion mutant proteins. Previous experiments had been carried out under conditions expected that M1 forms an aggregate. Therefore, we separated a soluble form from an aggregated form by fractionation through glycerol density gradient. Then, we systematically examined the effect of M1 on various steps of viral RNA synthesis, i.e., RNA binding, cleavage of the capped RNA, incorporation of the first nucleotide, and elongation of RNA chains.

First, the RNA binding activity of M1 was examined using the RNA mobility shift assay with a single-stranded RNA probe of 172 bases (24). At the low concentration of M1, a fast

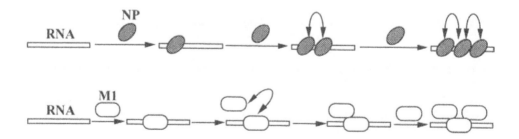

Figure 7. Cooperative RNA binding of NP and M1. An NP molecule binds to an RNA, and the second NP binds to the RNA cooperatively interacting with the first NP on the RNA. Alternatively, a NP-NP interaction prior to RNA binding may facilitate RNA binding of the NP-NP complex more effectively than that of a NP. A M1 molecule binds to an RNA, and the second M1 recognizes to the M1 on the RNA rather than an RNA.

migrating band corresponding to an intermediate M1-RNA complex was formed, while at the increasing concentrations, large M1-RNA complexes appeared without formation of different kinds of intermediate complexes. This could be explained by the concept that M1 binds to RNA and then a free form of M1 preferentially binds to this M1-RNA complex rather than free RNA by M1-M1 cooperative interaction (Figure 7). On this line, it has been reported that influenza virus NP and adenovirus DNA binding protein, both of which are cooperatively bound to RNA and single-stranded DNA, respectively, form the highly associated complex without any intermediate complexes (7, 25). Furthermore, analyses using glycerol density gradient revealed that M1 is associated with RNA-NP complexes, forming complexes similar to the huge M1-RNP complexes from virions in size. These observations lead the hypothesis that in the presence of large amount of M1, that is, in cytoplasm of infected cells at late stages, RNP is covered with M1 due to cooperative binding of M1 to RNP, and then readily incorporated into envelope.

Next, we examined the effect of His-M1 on *cell-free* viral RNA synthesis (24). M1 caused approximately 50% inhibition in RNA synthesis from the endogenous template, while RNA synthesis from exogenously added RNA template was much more efficiently inhibited than that from endogenous RNP. Therefore, it was indicated that M1 is capable of inhibiting viral RNA synthesis as well as binding to RNA/RNP. The M1-mediated inhibition of RNA synthesis was found to operate from its early stages. Transcription initiates with cleavage of capped RNA, followed by incorporation of GMP into the 3' end of the cleaved capped RNA. Since His-M1 inhibited this reaction process, M1 would inhibit cleavage reaction and/or GMP incorporation.

Finally, we mapped functional domains of M1 using a series of mutant recombinant His-M1 proteins. The experiments with M1 mutant proteins containing amino acid deletion suggested that the M1 region between amino acid residues 91-111 was essential for RNA binding, inhibition of capped RNA cleavage, and anti-RNA synthesis activity. It is therefore presumed that M1 is bound to RNA or RNA on the RNP complex, thereby inhibiting the initiation of the RNA synthesis. The arginine and lysine-rich region spanning amino acids 95-105 was suggested to interact with the phosphate backbone of RNP (26). It was reported that anti-RNA synthesis and RNA binding domain were localized at the region between amino acid residues 128-164 and/or 90-108 (27, 28). These contradictory results may be obtained due to different experimental conditions. Previous experiments were performed under the low salt, whereas our experiments were performed under the physiological salt concentrations. Indeed, a mutant M1 protein lacking amino acid residues 91-111, that was incapable of binding to RNA at the physiological conditions, showed the low but distinct RNA binding activity at the low salt concentrations (data not shown). Thus, it is likely that the region between amino acid positions 128-165 may form the RNA binding domain which is less effective in RNA binding and transcription inhibition than the region between amino acid positions 91-111. As wild type His-M1 and mutant M1 proteins that are bound to RNA formed highly associated complexes in the RNA mobility shift assay, the same M1 region turned out to be involved in oligomerization of M1 on RNA. Thus, in our assay the RNA binding and the oligomerization domains were not separable. It is noted that the region determined in our analyses contains the signal responsible for nuclear localization spanning amino acid residues between 101-105 (29). M1 is found transported into nucleus, so that it may be involved in inhibition of initiation and/or re-initiation of replication and secondary transcription at late infection stages and in the switch from the RNA synthesis mode to virion assembly.

3.3. Nuclear export of viral components

Transcription and replication of the influenza virus genome occur in nucleus. The incoming vRNPs support the primary transcription leading synthesis of early gene products, PB1, PB2, PA, NP, and NS1. The secondary transcription occurring after onset of the genome replication generates mRNA for late gene products, HA, NA, M1, M2, and NS2. These proteins are essential for the formation of a mature virus particle, which occurs in cytoplasm.

Of the viral components, vRNP complexes, NP, and M1 are thought to shuttle between the nucleus and cytoplasm through the nuclear pore complex (NPC) (30 for review). Importin α, a subunit of Importin binds to the nuclear localization signal (NLS) and Importin β mediates docking of the NLS-Importin complex with NPC (31 for review). A yeast two hybrid system identified two Importin α family proteins, designated NPI-1 and NPI-3, that interact with NP (32, 33). The nuclear export signal (NES) is identified as signal for protein nuclear export (34 for review). CRM1/Exportin1 is a receptor for proteins containing NESs (35). Binding of CRM1/Exportin1 to the NES is sensitive to the cytotoxin, leptomycin B (LMB) (36). Since incoming vRNP is imported to nucleus and newly synthesized progeny vRNP is exported from nucleus, a proper regulatory mechanism(s) should operate during these processes (37 for review).

Although it is suggested that M1 and NS2, a M1 binding protein, play roles in export of vRNP (38, 39), the export mechanism remains controversial. We tried to elucidate the mechanism of nuclear export of vRNP complexes during the late stages of infection using digitonin-permeabilized MDCK cells and LMB.

To biochemically study the mechanism of nuclear export of RNP complexes, we intended to utilize digitonin-permeabilized cells that have been used for the *in vitro* study of the nuclear import mechanism (40). When cells are permeabilized with a low concentration of digitonin, a nonionic detergent, the plasma membrane is selectively perforated while the nuclear envelope and other major membrane organelles are kept intact and cytosolic components are released from the cells. In our hands, permeabilized cells were correctly prepared, since FITC-conjugated recombinant NP and M1 proteins were properly imported into nucleus (22). We then used these permeabilized cells to study the nuclear export of viral components (22). Influenza virus-infected MDCK cells at late stages were permeabilized by treatment with digitonin at $0°$ C and washed successively with a buffer to be used for nuclear export to remove soluble cytoplasmic components such as viral proteins and nuclear import factors. The export of viral RNA polymerases, NP and NS2, those of which had been accumulated in nuclei, was observed in an incubation period-dependent manner when incubation was carried out at $30°$ C but not at $0°$ C, while cellular nuclear proteins stayed in nuclei under any conditions used. In addition, exported fractions were RNA-synthesis competent, suggesting that the RNA polymerases and NP are not present as free forms but forming RNP complexes. However, M1 came out from nuclei even at $0°$ C. Small molecules less than 40 kilodaltons (kDa), readily diffuse through NPC. Therefore, it is possible that M1 (26 kDa) is easily dissociated from its oligomeric form and/or from its nuclear binding target and NS2 (14 kDa) may be associated with a molecule/complex with a large molecular mass.   These observations also indicate the pre-existing factors in nuclei could be sufficient for the export of viral proteins.

A viral late protein, NS2 containing a putative NES, is suggested to promote the export of vRNP (39).   Then we examined the effect of LMB on the export of viral components. Western blotting analyses of exported proteins revealed that he nuclear export of NP and the RNA polymerases was significantly inhibited by the LMB, while LMB treatment did not significantly affect the level of NS2 export.   These results suggested that the export of NP and the RNA polymerases seems regulated by a factor(s) affected by LMB and NS2 may not be involved in vRNP export.   This concept disagreed with the proposal that NS2 mediates nuclear export of vRNP via its NES (39) and via its binding capability to M1 (41, 42, and our unpublished data).   These contradictory results can be explained by a hypothesis that there exist two types of NS2, i.e., a LMB-sensitive NS2 which is transiently and/or weakly associated with vRNP and a LMB-insensitive NS2 which is not associated with vRNP.   In addition, a major fraction of M1 is exported from nucleus regardless of LMB treatment. This suggested that M1 is most likely to be either exported independent of a LMB-sensitive pathway and/or a vRNP export pathway or diffusely exit nuclei.

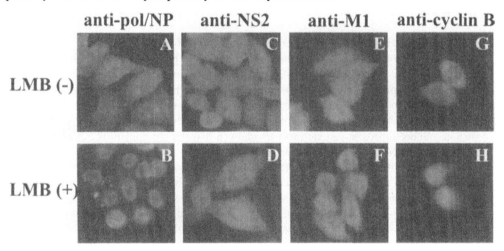

Figure 8.   Effects of LMB on localization of NP/Polymerases, M1 and NS2.
HeLa cells infected with influenza virus A/PR/8/34 at M.O.I. of 5 were incubated for 5 hr, and then treated with or without 10 ng/ml LMB for 2 hr.   Cells were fixed and stained with anti-polymerase/NP antibody (panels A and B), anti-NS2 antibody (panels C and D), and anti-M1 antibody (panels E and F).   As a control, mock-infected HeLa cells treated with or without 10 ng/ml LMB for 2 hr were stained with anti-cyclin B antibody (panels G and H).

To confirm the intracellular localization of vRNP, M1 and NS2 in the presence of LMB, indirect immunofluorescence assays were performed.   At 7-hour post infection, the viral RNA polymerases and/or NP are observed in both cytoplasm and nucleus (Figure 8A). When cells were treated with LMB, they accumulate exclusively in nucleus (Figure 8B). NS2 and M1 are localized in cytoplasm and nucleus (Figures 8C and 8E), while intracellular localization of NS2 or M1 in the presence of LMB is not significantly different from that in the absence of LMB (Figures 8D and 8F). These results suggested that the nuclear export of RNP, but neither M1 nor NS2, is dependent on a CRM1/Exportin1-dependent nuclear export pathway. Further, the vRNP export is unlikely to be dependent on the localization pattern of M1 and NS2. When exported fractions were subjected to the glycerol density gradient, the RNA polymerases and NP sedimented in fractions corresponding to vRNP complexes, while M1 and NS2 were recovered in top fractions of the gradient. These results suggested that major fractions of M1 and NS2 are exported form nucleus without associating to vRNP complexes. Although M1 did not seem directly involved in vRNP export, it is still possible that it interacts with vRNP and/or nuclear components, such that vRNP is ready to exit nucleus. Alternatively, it must not be overlooked that a small fraction of M1 that is hardly detected in our assays, could also be involved in the vRNP export.

In conclusion, results in these experiments altogether suggested that nuclear export of RNA synthesis-competent vRNP is dependent on a LMB-sensitive pathway but independent on NS2 and M1. However, we could not rule out the possibility that M1 and NS2 are exported through both CRM1/Exportin1-dependent and -independent pathways, but that in the presence of LMB the CRM1/Exportin1-independent pathway dominates.

3.4. Interaction between RNP-M1 complexes and membrane components

After the genome replication in nuclei and the nuclear export of progeny RNP complexes, the exported RNP complexes are incorporated into progeny virions in cytoplasm. It has been widely accepted that viral transmembrane glycoproteins, HA and NA, are essential for virus entry and budding process. Cytoplasmic tail of HA and NA are conserved in various virus strain, suggesting that cytoplasmic tails of these glycoprotein are important for some function. Indeed, reverse-genetic methods showed that the tail domain is not essential but affects the efficiency of virion production (43, 44).

M1 is thought to interact with both RNP complexes and membrane fractions and be involved in virion formation. As described above, M1 is a multifunctional protein, thereby consisting of distinct functional domains. We have indicated that the region between amino acid residues 91-111 is involved in RNA binding, transcription inhibition, and self-oligomerization. Recently, amino acid residues between 101-105 have been shown required for nuclear localization (29). There are three hydrophobic domains in the primary structure of M1 and the hydrophobic regions between amino acid residues 62-68 and 114-133 are shown to be lipid binding domains (8).

Although M1 is suggested to play a key role for virion assembly, its function in the virion assembly mechanism is yet uncertain. To get a cue, we tried to examine membrane and/or HA binding properties of M1 using purified recombinant M1 proteins. Here, we summarize the outline of our results, since details for experimental methods and results will be described elsewhere.

The simple question was where M1, if it does, interacts with the viral glycoproteins. It has been generally believed that HA and NA are transported via vesicular transport from the ER through the Golgi apparatus to the plasma membrane. Using [^{35}S]-methionine labeled infected MDCK cells, we tested whether M1 is associated with *cis*-Golgi in the presence of BFA, an inhibitor for transport from ER to the *cis*-Golgi, and under low-temperature conditions (at 20° C) that inhibits transport from the *trans*-Golgi to the plasma membrane, respectively.   These treatments accumulated HA on *cis*-Golgi and *trans*-Golgi, respectively, but did not M1 on them, suggesting that M1-HA association occurs after HA is transported to the *trans*-Golgi network.

Cell-free pull-down assays using Ni-chelation resins revealed that histidine-tagged recombinant M1 interacted with viral membrane fractions isolated from virions including HA. This was confirmed in experiments using recombinant M1 and HA proteins.   Furthermore, M1 was found to interact with synthetic liposome vesicles.   Using a series of deletion mutants of M1 proteins, it was suggested that bipartite regions outside the RNA binding domain were important for efficient membrane binding.

4. PERSPECTIVE

Here we have described our recent achievement on molecular mechanisms related to replication and transcription of the influenza virus genome and primary steps of virion assembly.   However, there still remains a variety of problems to be answered.   It is not yet clarified how the viral genome replication initiates, how the synthesis level of each progeny segmented genome RNA is coordinated, how the level of transcription from each segmented genome RNA is regulated, what is a switching mechanism between replication and transcription modes, where replication and transcription take place in nucleus, and *etc.*   It is interesting and important to study how premature genome RNAs are not incorporated into virions, how a set of the segmented genome RNAs is sorted and assembled, how the envelope structure is determined, and *etc.*

We have been dissecting and reconstituting the systems that reproduce reactions found in infected cells.   Recently, in addition to infectious RNP systems, a novel system for generation of recombinant viruses entirely from its cDNA were reported (11).   These systems enable to make various mutant viruses that are to be quit useful for genetic analyses. However, it is difficult by this system to generate recombinant viruses that are lethal.   Thus, we believe that the biochemical ways, by which we can identify and study factors and mechanisms involved in processes of interest, are useful and complementary for genetic approaches.   With these methods, we hope not only to gain insight into the fundamental mechanisms but also to devise a novel design(s) to control the virus disease.

5. ACKNOWLEDGEMENTS

I acknowledge people for their helpful collaboration, in particular, K. Mizumoto (Kitasato University), A. Ishihama (National Institute of Genetics), S.Nakada (Science University of Tokyo), P. Palese (Mount Sinai School of Medicine), A. Iwamatsu (Kirin Brewery Company), N. Kobayashi (Nagasaki University), T. Turan (University of Istanbul), K. Yamanaka (Kumamoto University), H. Mihara, T. Sato, K.Shimizu, K. Nakade, K. Hoshida, K. Watanabe, F. Momose, M. Mibayashi, M. Katoh, K. Sugiyama, N. Takizawa, A. Numajiri, T. Naito, and K. Pepin (Tokyo Institute of Technology).

Research from our laboratory was supported by a grant from the Biodesign Research Project of RIKEN, grants from Ministry of Education, Culture, Sports, Science, and

138

Technology of Japan, a grant-in-aid from the Tokyo biochemical Research Foundation, a KAST grant, and a grant-in-aid from Asahi Glass Foundation.

REFERENCES

1. U. Desselberger, V. R. Racaniello, J. J. Zazra, P. Palese, Gene, 8 (1980) 315.
2. E. Fodor, D. C. Pritlove, and G. G. Brownlee, J. Virol., 69 (1995) 4012.
3. H. J. Kim, E. Fodor, G. G. Brownlee, and B. L. Seong, J. Gen. Virol., 78 (1997) 353.
4. R. Flick and G. Hobom, J. Gen. Virol., 80 (1999) 2565.
5. D. C. Pritlove, L. L. Poon, L. J. Devenish, M. B. Leahy, and G. G. Brownlee, J. Virol, 73 (1999) 2109.
6. K. Nagata, F. Momose, and M. Okuwaki, Recent Res. Devel. Virol., 1 (1999) 559.
7. K. Yamanaka, Ishihama, A., and K. Nagata, J. Biol. Chem., 265 (1990) 11151.
8. A. Gregoriades and B. Frangione, J. Virol., 40 (1981) 323.
9. W. Luytjes, Krystal M., M. Enami, J. D. Parvin, and P. Palese, Cell, 59 (1989) 1107.
10. K. Yamanaka, N. Ogasawara, H. Yoshikawa, A. Ishihama, and K. Nagata, Proc. Natl. Acad. Sci. USA, 88 (1991) 5369.
11. G. Neumann, T. Watanabe, H. Ito, S. Watanabe, H. Goto, P. Gao, M. Hughes, D. R. Perez, R. Donis, E. Hoffmann, G. Hobom, and Y. Kawaoka, Proc. Natl. Acad. Sci. USA, 96 (1999) 9345.
12. K. Nagata, K. Takeuchi, and A. Ishihama, J. Biochem., 106 (1989) 205.
13. K. Shimizu, H. Handa, S. Nakada, and K. Nagata, Nucleic Acids Res., 22 (1994) 5047.
14. J. D. Dignam, R. M. Lebovitz, and R. G. Roeder, Nucleic Acids Res., 11 (1983) 1475.
15. F. Momose, H. Handa, and K. Nagata, Biochimie, 78 (1996) 1103.
16. F. Momose, C. F. Basler, R. E. O'Neill, A. Iwamatsu, P. Palese, and K. Nagata, J. Virol., 75 (2001) 1899.
17. J. Fleckner, M. Zhang, J. Valcarcel, and M. R. Green, Genes Dev., 11 (1997) 1864.
18. J. P. Staley and C. Gutrhrie, Cell, 92 (1998) 315.
19. Z. Chen, Y. Li, and R. M. Krug, EMBO J., 18 (1999) 2273.
20. T. Wolff, R. E. O'Neill, and P. Palese, J. Virol., 72 (1998) 7170.
21. O. M. Rochovansky and G. K. Hirst, Virology, 73 (1976) 339.
22. K. Watanabe, N. Takizawa, M. Katoh, K. Hoshida, N. Kobayashi, and K. Nagata, Virus Res., 77 (2001) 31.
23. K. Yamanaka, A. Ishihama, and K. Nagata, Virus Genes, 2 (1988) 19.
24. K. Watanabe, H. Handa, K. Mizumoto, and K. Nagata, J. Virol., 70 (1996) 241.
25. P. A. Tucker, D. Tsernoglou, A. D. Tucker, F. E. J. Coenjaerts, H. Leenders, and P. C. van der Vliet, EMBO J., 13 (1994) 2994.
26. G. Winter and S. Fields, Nucleic Acids Res., 8 (1980) 1965.
27. Z. Ye, R. Pal, J. W. Fox, and R. R. Wagner, J. Virol., 61 (1987) 239.
28. Z. Ye, N. W. Baylor, and R. R. Wagner, J. Virol., 63 (1989) 3586.
29. Z. Ye, D. Robinson, and R. R. Wagner, J. Virol., 69 (1995) 1964.
30. M. Ohno, M. Fornerd, and I. W. Mattaj, Cell, (1998) 327.
31. D. Gorlich and I. W. Mattaj, Science, 271 (1996) 1513.
32. R. E. O'Neill, S. R. Jaskunas, G. Blobel, P. Palese, and J. Moroianu, J. Biol. Chem., 270

(1995) 22701.

33. R. E. O'Neill and P. Palese, Virology, 206 (1995) 116.

34. L. Gerace, Cell, 82 (1995) 341.

35. M. Fornerod, M. Ohno, M. Yoshida, and I. W. Mattaj, Cell, 90 (1997) 1051.

36. M. Fukuda, S. Asano, T. Nakamura, M. Adachi, M. Yoshida, M. Yanagida, and E. Nishida, Nature, 390 (1997) 307.

37. G. R. Whittaker and A. Helenius, Virology, 246 (1998) 1.

38. G. R. Whittaker, M. Bui, and A. Helenius, Trends Cell Biol., 6(1996) 67.

39. R. E. O'Neill, T. Julie, and P. Palese, EMBO J., 17 (1998) 288.

40. S. Adam, R. S. Marr, and L. Gerace, J. Cell. Biol., 111 (1990) 807.

41. J. Yasuda, S. Nakada, A. Kato, T. Toyoda, and A. Ishihama, Virology, 196 (1993) 249.

42. A. C. Ward, L. A. Castelli, A. Lucantoni, J. F. White, A. A. Azad, and I. G. Macreadie, Arch. Virol., 140 91995) 2067.

43. H. Jin, G. P. Leser, and R. A. Lamb, EMBO J., 13 (1994) 5504.

44. A. Garcia-Sastre and P. Palese, Virus Res., 37 (1995) 37.

Molecular Anatomy of Cellular Systems
I. Endo et al., (editors)
© 2002 Elsevier Science B.V. All rights reserved.

Mechanisms of regulation of eukaryotic homologous DNA recombination

Takehiko Shibata [a, b], Ken-ichi Mizuno [b, c] and Kunihiro Ohta [b, c]

[a] Cellular & Molecular Biology Laboratory, RIKEN (The Institute of Physical and Chemical Research)
[b] CREST, Japan Science and Technology Corporation (JST)
[c] Genetic Dynamics Research Unit-Laboratory, RIKEN

Homologous genetic (or DNA) recombination is an exchange or replacement of a segment of parental DNA with a segment having the identical or very similar (*i.e.*, homologous) sequence from a partner DNA. This process is essential for the repair of double-stranded DNA breaks that occur during normal DNA replication or by DNA damaging agents. In eukaryotes, homologous recombination is essential to meiosis, and at an early stage of meiosis homologous recombination is transiently induced to a frequency several orders of magnitude higher than that of mitotic cells. How homologous recombination is regulated is still unclear. We found that around the initiation sites of homologous recombination, two different kinds of the changes in chromatin structures play important roles in the regulation of homologous recombination: site-specific changes (chromatin transition) and extensive changes in a region of hundreds of base pairs (chromatin remodeling). The chromatin transition was detected in both *Saccharomyces cerevisiae* and *Schizosaccharomyces pombe* as a time- and site-specific transient increase in the hypersensitivity of chromatin DNA towards micrococcal nuclease (MNase). In *S. cerevisiae*, the transition, which precedes the meiotic double-stranded DNA breakage, occurs predominantly in promoter regions at a subset of constitutive nuclease hypersensitive sites that correlate well with DNA breakage sites. The transition depends on the function of *MRE11*. The Mre11 protein has two functionally independent domains, one of which (the C-terminal domain) plays a critical role in chromatin transition. Chromatin remodeling occurs during sporulation-cultivation around the *M26* mutation site in the *ade6* locus, a well-known hot spot of meiotic recombination. The remodeling depends on an ATF/CREB-family transcription factor and *cis*-acting sequences belonging to the CRE family, and creates MNase-hypersensitive sites where chromatin transition occurs during meiosis. Thus, *S. pombe* and probably also higher eukaryotes have a hierarchical mechanism to regulate the initiation of meiotic recombination at the chromatin level.

1. INTRODUCTION

Genomic DNA appears to duplicate precisely for the transmission of a complete copy of genomic information to the descendant. However, several kinds of rearrangements of genomic DNA, called genetic or DNA recombination, have been observed in various organisms; these rearrangements include crossing over of homologous chromosomes, translocation of chromatin segments, DNA-mediated transformation, and sister chromatid exchange. Until recently, it was generally believed that DNA rearrangements were rare and stochastic events or induced in response to DNA damage.

Homologous recombination is a type of genetic recombination in which a segment of a parental DNA is exchanged with ("crossing over") or replaced by ("gene conversion") a homologous segment of partner DNA (Fig. 1). Since the first group of mutants deficient in homologous recombination was isolated in the mid-60's from the bacterium *Escherichia coli* [1], various other mutants have been isolated or identified from bacteria, simple eukaryotes such as yeasts (*Saccharomyces cerevisiae, Schizosaccharomyces pombe*) and fungi, as well as higher eukaryotes. More recently, some inherited diseases in humans were found to involve genetic recombination [2, 3]. Studies on the homologous recombination-deficient mutants revealed that homologous recombination is essential to both DNA repair and meiosis.

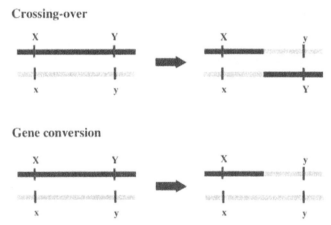

Fig. 1   Crossing over and gene conversion.   Each bar represents double-stranded DNA.   X and x, and Y and y represent alleles at X and Y loci, respectively.

Except in some viruses, genomic DNA is double stranded, and damage to the bases, sugars, or the strand backbone are repaired by copying the partner strand. However, to repair double-stranded DNA-breaks, an intact intramolecular template strand is not available; therefore, homologous recombination is essential to accurately repair this type of DNA damage.

Recent studies on the function of genes in higher eukaryotes homologous to yeast genes specific to homologous recombination showed that homologous recombination is essential to cell proliferation, development, and differentiation. Replication forks are collapsed when the forks encounter a single-stranded break during DNA replication. The collapsed replication fork is restored by the interaction of the nascent DNA terminus with the sister chromatid through the same process to initiate homologous recombination [4-6] (see ref. [7] for review).

It has been recognized that some prokaryotes and most eukaryotes have systems to regulate homologous recombination in a programmed manner. Under starvation or high cell density, cells of the prokaryote *Bacillus subtilis* activate machinery to incorporate extracellular DNA from the same species into their chromosomal DNA [8, 9]. In eukaryotes, homologous recombination is an integral part of sexual reproduction. Homologous recombination is essential to meiosis at the disjunction of homologous chromosomes (sorting of a pair of homologues into a pair of sister cells) and induced transiently at an early stage of meiosis to several orders of magnitude higher than it is in mitotic cells (see refs. [10-14] for review). The initiation of meiotic homologous recombination is mediated by programmed site-specific double-stranded breaks [15-18]. How homologous recombination is regulated is unclear.

To understand the regulation of homologous recombination, meiotic cells are the best targets for research since the clear and extensive induction of homologous recombination occurs at a very limited stage. Meiosis and thus meiotic homologous recombination can be induced synchronously in the budding yeast *S. cerevisiae*, the fission yeast *S. pombe*, and the mushroom *Coprinus cinereus* under laboratory conditions. Until recently, no other organisms were available for these types of experiments. In addition, both yeasts were extensively investigated in genetic and molecular studies; consequently, various mutants relating to homologous recombination and meiosis are available. Thus, we chose both yeasts to study the role of chromatin structures in the regulation of homologous recombination. In this article, we will describe the background of our research on the regulation of meiotic homologous recombination at the chromatin level and discuss our conclusion that two levels of hierarchical chromatin structural changes play a pivotal role in regulating the initiation of meiotic recombination.

2. HOMOLOGOUS RECOMBINATION

2.1. Genetic mapping and recombination hot spot

DNA recombination is classified into four groups, depending on how sites of recombination are selected. Site-specific recombination occurs between a pair of strictly specified DNA sequences, while transposition occurs between a specified sequence and a less specified sequence. Non-homologous or illegitimate recombination occurs between DNA molecules with no or very short sequence homology (one to a few base pairs). Homologous recombination occurs between a pair of DNA segments with homologous sequences from tens to hundreds of base pairs in length, such as recombination between homologous chromosomes and that between sister chromatids. Homologous DNA recombination was first recognized as a crossing over process and was proposed to be a measure for the distances between genes by T. H. Morgan in 1912 [19]. Since then, genetic maps of various organisms have been successfully constructed by assuming that crossing over occurs between homologous chromosomes at a constant frequency anywhere along the chromosomes and that the frequency of crossing over is proportional to the distance between genes. The centimorgan (cM), a unit of genetic distance (giving 1% recombination), was termed in his honor. It is likely that this work gave rise to the belief that homologous recombination is a stochastic event. Today, however, we know that recombination frequency is not random throughout the whole chromosome; *i.e.*, chromosomes have numerous recombination hot spots where the frequency of recombination is much higher than in neighboring areas. Moreover, homologous recombination is transiently and specifically induced at an early stage of the first meiotic cell division, and at the sites of DNA lesions, especially double-stranded breaks caused by DNA damaging agents in mitotic cells. The meiotic recombination hot spots are the sites of the initiation of homologous recombination in meiosis. In *S. cerevisiae*, it was demonstrated that at the early stage of meiosis, well-regulated double-stranded breaks are transiently introduced at the hot spots to induce homologous recombination between homologous chromosomes.

2.2. The Holliday intermediate, a common intermediate for crossing over and gene conversion

As described in the introduction, there are two types of homologous recombination: crossing over and gene conversion (see Fig. 1). Crossing over is explained by the breakage and rejoining of DNA, while gene conversion appears to be the result of gene copying. It is well known that crossing over between a pair of linked genes flanking a gene that was converted is always about 50% (no apparent linkage). This constant relationship between crossing over and gene conversion suggests a single common mechanism for both classes of homologous recombination. When this relationship was recognized, scientists preferred the copy choice (or replicative template switch) model to the breakage-and-rejoining model as the mechanism. However, biophysical studies on the recombination process and its products indicated that recombination occurs even in the absence of DNA replication and that recombinant DNA is a physical chimera of parental DNA molecules [20-22]. Some scientists proposed hybrid DNA (heteroduplex joints) as intermediates of homologous recombination, with each strand derived from each parental DNA. Joining of parental DNA molecules by heteroduplex formation and the resolution of the joint explains crossing over, while the mismatch repair within the heteroduplex results in gene conversion. The Holliday intermediate, a type of recombination intermediate containing a pair of heteroduplexes, was proposed by R. Holliday [23] and was shown to be present in the prokaryote *E. coli* [24] and in eukaryotic cells (yeast) undergoing homologous recombination [25].

The *E. coli* RecA protein was found to have an ATP-dependent activity to form heteroduplex joints *in vitro* from homologous double-stranded and single-stranded DNA [26, 27]. The RecA protein alone could form a Holliday intermediate from a pair of homologous linear double-stranded molecules if one of them had a single-stranded tail(s). Subsequently, it was revealed that yeasts [28-32], a fungus [33], and mammals [32] have a family of proteins (RecA/Rad51-family) that are homologues of the RecA protein and that some of them can form a heteroduplex joint *in vitro* [34]. The similarity between the three-dimensional structure of the Rad51 protein-DNA complex and the RecA protein-DNA complex are much more impressive than their homology at the amino acid level [35]. In *S. cerevisiae*, RAD51, the gene encoding the Rad51 protein, is one of the genes essential to both meiotic homologous recombination and DNA repair of double-stranded breaks in mitotic cells.

2.3. Double-Strand break-repair model

The homologous integration of exogenous DNA into a chromosome of *S. cerevisiae* was extensively enhanced by *RAD52*, a gene critical for meiotic recombination, when exogenous DNA has a double-stranded break or a gap within a nucleotide sequence homologous to that of the recipient cell's chromosomal DNA [36].   This finding and the heteroduplex joint formation by the RecA protein inspired Szostak *et al.* to propose a double-strand break-repair model as a mechanism of meiotic homologous recombination [15].   A series of studies by Szostak and his collaborators support the double-stranded break-repair model; these include *cis*-acting sequences within the promoter of ARG4 [37], temporal double-stranded DNA breakage within the sequence [16], single-stranded tail formation by the processing of the double-stranded break [16], and the physical detection of the formation of a heteroduplex at this site [38].

3. *cis*-ACTING SEQUENCES AND *trans*-ACTING FACTORS THAT REGULATE MEIOTIC HOMOLOGOUS RECOMBINAITON

3.1. Hot spot for homologous recombination: *cis*-acting sequences and *trans*-acting factors

A hot spot of meiotic recombination was first found at the his-3 locus in *Neurospora crassa* [39] and then at the *ade6* locus in *S. pombe* [40].   These recombination hot spots are created by mutations and act as the initiation sites for meiotic recombination.   The hot spot at the *ade6* locus created by the *M26* mutation enhances meiotic recombination by an order of magnitude.   The *M26* mutation is a single base substitution that creates the heptanucleotide sequence ATGACGT [41, 42].   When the *ade6* locus is heterozygous to the *M26* mutation, meiotic recombination is enhanced at the locus, but the majority of the progenies do not contain the *M26* mutation because of a strong polar gene conversion event at this locus.   This is a general consequence of gene conversion at a heterozygous locus, only one of which is a recombination hot spot.   The *M26* hot spot enhances not only gene conversion but also crossing over [43].

The binding of the heterodimeric transcription factor Atf1-Pcr1 to the heptameric sequence ATGACGT is required for the hot spot activity of the *M26* mutation [44, 45].   Atf1-Pcr1 is a member of the ATF/CREB-family, which are cyclic-AMP-responsive transcription factors, and the heptameric sequence resembles the consensus sequence of the transcriptional *cis*-acting sequences found in the promoter region of genes whose expression is regulated by the levels of cyclic-AMP (cyclic-AMP-responsive element or CRE) [46].   Atf1-Pcr1 plays a crucial role in the induction of meiosis and the response to environmental stresses [45, 47-49], suggesting that homologous recombination by itself is an important response mechanism to environmental stresses.

Extensive studies on gene conversion at the *ARG4* locus of *S. cerevisiae* revealed a hot spot for initiation of meiotic homologous recombination.   Genetic studies revealed a polar gradient in the frequencies of gene conversion within this and other loci; *i.e.*, the frequency is higher for genetic markers proximal to the promoter and lower for distal markers.   These observations suggest the presence of *cis*-acting sequences for meiotic recombination in the promoter regions [50].

A *cis*-acting sequence in the promoter region of the *ARG4* locus of *S. cerevisiae* was identified and characterized [37, 51]. Transient double-stranded breaks were introduced within a limited region of the *cis*-acting sequence at the early stage of meiosis, as described. The modifications of core sequences revealed a positive correlation between the extent of double-stranded breakage and the frequency of gene conversion for a genetic marker in the *ARG4* locus, supporting a model that meiotic recombination is initiated by double-stranded breaks in this *cis*-acting sequence [51].

3.2. Genes involved in meiosis-specific double-stranded DNA breakage

Genetic studies identified the *MRE11, RAD50, XRS2*, and *SPO11* genes as essential for meiotic double-stranded DNA breakage and the initiation of homologous recombination. The Rad50 protein is required for double-stranded DNA breakage as well as for the processing of the nascent termini to single-stranded tails at the site of double-stranded breaks. This conclusion is based on the observation that in cells with a partial reversion mutation, *rad50S*, meiotic double-stranded breaks are introduced at the early stage of meiosis but are not processed or repaired. As described later, the Mre11 protein also has similar dual functions of double-stranded DNA breakage and processing of the double-stranded breaks. The Spo11 protein very likely contains the endonuclease activity required to introduce double-stranded DNA breaks since this protein is covalently attached to the DNA termini of cells with the *rad50S* mutation [52] and the amino acid sequence of the Spo11 protein is homologous to a type II topoisomerase of an archaea [53]. It is well known that type II topoisomerases introduce transient double-stranded breaks during the process to change supercoiling and that at these transient breaks, the topoisomerase attaches one of the termini with a covalent bond.

4. MEIOSIS-SPECIFIC CHROMATIN TRANSITION AT THE INITIATION SITES FOR MEIOTIC RECOMBINATION

4.1. Meiotic double-stranded DNA breakage and nuclease-hypersensitive sites

Most meiotic double-stranded breaks for the initiation of meiotic homologous recombination are introduced in the transcriptional promoter regions of *S. cerevisiae* [18]. Both Lichten's group and our group found that the double-stranded break sites exhibited DNase I- and micrococcal nuclease (MNase)-hypersensitivity in chromatin from meiotic and mitotic cells [54, 55]. Using gel electrophoresis, the *ARG4* and *CYC3* loci were shown to contain one and two double-stranded break sites, respectively. Each of the double-stranded break sites detected by agarose gel electrophoresis were shown to contain some double-stranded break sites when they were analyzed at the single nucleotide resolution [56]. We examined the chromatin structures around the meiotic double-stranded break sites in the *ARG4* and *CYC3* loci of *S. cerevisiae* cells that were committed to meiosis. In order to probe chromatin structure, we digested chromatin DNA isolated from meiotic cells with micrococcal nuclease (MNase), isolated the digested DNA, and redigested it with restriction endonucleases. The DNA digests were separated by agarose gel electrophoresis and probed by an indirect end labeling technique. MNase preferentially cleaves chromatin DNA at linker regions between nucleosomes, DNA regions with high accessibility to proteins by the absence of nucleosomes, and those with altered tertiary DNA structures. Using this procedure, we observed that all double-stranded break sites at both loci overlapped with a subset of MNase hypersensitive sites found in the promoter region [55]. The meiotic double-stranded break sites examined by Wu and Lichten and by us were all within the promoter region. Consequently, the observation that the meiotic double-stranded break sites are hypersensitive to nucleases is expected. However, by studying the chromatin structures around meiotic double-stranded DNA breaks sites in meiotic yeast cells, we obtained the unexpected results described in the following section.

4.2. Increased hypersensitivity towards micrococcal nuclease

We followed the fate of the MNase-hypersensitive sites at the meiotic double-stranded break sites and other MNase-hypersensitive sites in the same loci during the progression of meiosis. We found that at all MNase-hypersensitive sites that overlapped the meiotic double-stranded break sites, MNase-hypersensitivity increased 2- to 4-fold prior to double-stranded DNA breakage and the appearance of recombinant DNA. This increase in MNase sensitivity is specific to sites overlapping the double-stranded break sites [55]. We compared the MNase hypersensitivities of chromatin DNA with modified *cis*-acting sequences causing either hyper- or hypo-recombination at the *ARG4* locus. The results exhibited that the extent of MNase hypersensitivity correlated well with the extent of meiotic double-stranded DNA breakage and the frequency of gene conversion at that locus [55]. These results suggest that local chromatin structures and their modification in an early stage of meiosis play an important role in the positioning and frequency of double-stranded DNA breakage leading to the initiation of meiotic homologous recombination. Such an increase in MNase sensitivity at the initiation site for meiotic recombination during an early stage of meiosis was also observed in *S. pombe* [57]. Since *Saccharomyces* and *Schizosaccharomyces* are phylogenetically unrelated, the increase in MNase hypersensitivity at the initiation sites for meiotic recombination reflects a general step towards the initiation of recombination. The increase in hypersensitivity at meiotic double-stranded break sites was not detected by DNaseI, suggesting that it was not simply due to a general increase in the accessibility of chromatin to proteins.

4.3. Genes required for the increase in micrococcal nuclease hypersensitivity at the initiation sites for meiotic homologous recombination

Among the genes required for meiotic double-stranded DNA breakage in *S. cerevisiae* described above (*MRE11, RAD50, XRS2,* and *SPO11*), only *MRE11* was found to be required for the increase in MNase-hypersensitivity at the double-stranded break sites [58]. It was described that the Rad50, Mre11, and Xrs2 proteins form a complex [59]. Thus, Mre11 protein plays a role to modify chromatin structure at the meiotic double-stranded break site independent of the Rad50 and Xrs2 proteins, and then Rad50, Xrs2, and Spo11 proteins are recruited to the site to introduce double-stranded breaks.

5. MOLECULAR ANATOMY OF THE Mre11 PROTEIN

5.1. Molecular functions of the Mre11 protein

We purified the wild-type and several mutant Mre11 proteins and found that the wild-type Mre11 protein has a DNA binding activity and a Mn^{2+}-dependent nuclease activity [60]. Similar results were reported for the *S. cerevisiae* Mre11 protein [59, 61] and the human Mre11 protein [62, 63]. In this section, we focus on the *S. cerevisiae* Mre11 protein.

Purified *S. cerevisiae* Mre11 protein is a multimer. The Mre11 protein binds either double-stranded or single-stranded DNA with a preference for the double-stranded form [59-61]. Experiments with oligo DNA demonstrated that the affinity of the Mre11 protein for DNA depended on the length of the double-stranded region rather than on its structure [60]. A mutant form of Mre11 deleting 49 amino acid residues from the C-terminus (mre11ΔC49) was found to lack DNA binding activity, suggesting the presence of DNA binding sites in the C-terminal region [60]. Another DNA binding site was located in the central region of the Mre11 protein [59].

The N-terminal domain of the Mre11 protein is homologous at the amino acid level to the *E. coli* sbcD protein, and this sequence has a phosphoesterase motif [64, 65]. SbcD protein exhibits an ATP-dependent double-stranded DNA exonuclease activity and an ATP-independent single-stranded DNA endonuclease activity following binding to the SbcC protein (SbcCD nuclease) [65, 66]. This suggests that the Mre11 protein has nuclease activities. The Mre11 proteins exhibited endonuclease activity towards circular phage single-stranded DNA and exonuclease activity on linear double-stranded DNA. Closed or open circular double-stranded DNA was not cut by the Mre11 protein [60]. The Endonuclease activity of the Mre11 protein is not simply single-stranded DNA-specific but recognize single-stranded DNA with secondary structures, since single-stranded oligo DNA are not a good substrate. Oligo-DNAs (ca. 30-mers) containing blunt ends or a 5' overhang were substrates for Mre11 nuclease activity, while 3' overhangs were not [59, 60]. These observations suggest that the exonuclease activity of Mre11 protein has 3' to 5' directionality [59, 60]. The nuclease activities require Mn^{2+} [59-61], which could not be replaced by other divalent cations such as Mg^{2+}, Ni^{2+}, Ca^{2+}, Sr^{2+}, Co^{2+}, and Zn^{2+} [60]. Unlike the SbcCD nuclease, the nuclease activities of Mre11 do not require ATP.

The human Mre11 protein was reported to open hairpin loops [63]. However, bubbles (unpaired regions) of 3 to 6 nucleotides within double-stranded DNA were not recognized by the Mre11 protein [60]. Further studies on the specificity of the Mre11 nuclease activities are in progress.

Mutant Mre11 proteins with an amino acid-replacement within the conserved phophoesterase motif (mre11D16A, mre11-58 (H213Y, L225I), and mre11D56N) were all shown to be defective in nuclease activity [59-61]. Another mutant Mre11 protein deleting a DNA binding site located in the middle of the Mre11 polypeptide (mre11-6) is also defective in nuclease activity [59]. On the other hand, mutant Mre11 proteins deleting regions proximal to the C-terminus, mre11ΔC49 (deleting 49 amino acid residues from the C-terminus) and mre11-5 (deleting 136 amino acid residues from the C-terminus), are fully active as nucleases [59, 60]. This indicates that the DNA binding site in the middle of the polypeptide is required for nuclease activity, while the suggested C-terminal DNA binding site is not.

5.2. The functional domain for repair of double-stranded breaks

A group of mutants (*rad50S* and *mre11S*) proficient in the introduction of meiotic double-stranded breaks at the initiation sites for meiotic recombination but deficient in the processing and repair of meiotic and mitotic double-stranded breaks is known [17, 67]. Cells containing mutant Mre11 proteins defective in nuclease activity (*mre11D16A, mre11-58, mre11H125N*) were found to be members of this group [59-61]. The increase in MNase hypersensitivity at the meiotic double-stranded DNA break sites was unaffected in cells having the mre11D16A mutant protein [60]. Cells with either *mre11D16A* or *mre11-58* were shown to be as sensitive as the *mre11-null* mutation to MMS (methylemetane sulfonate) treatment, which causes DNA damage. This suggests that the nuclease activity of the Mre11 protein plays a role in the repair of meiotic double-stranded breaks and DNA damage introduced by MMS treatment, but its nuclease activities is not required for meiotic double-stranded DNA breakage. After meiotic double-stranded DNA breakage, the DNA termini at the break sites had extensive 3' overhanging single-stranded sequences [68]. The Mre11 protein has a 3' to 5' exonuclease activity; this directionality is the opposite of that required for the production of the 3' overhanging sequences. It is possible that the Mre11 protein cleans up the 5' ends of DNA strands by removing covalently attached proteins or products of DNA damage. This enabled exonuclease(s) to degrade strands from the 5' termini to produce the 3' overhanging sequences. Further studies are awaited.

We also found telomere shortening in cells with a mutation causing a deficiency in the nuclease activity (*mre11D16A*) [60] of Mre11, which has been shown to play a role in telomere maintenance [69, 70]. This suggests that the nuclease activity of Mre11 is important for telomere maintenance in mitotic cells.

5.3. The functional domain for meiotic double-stranded DNA breakage

Cells with a mutant Mre11 protein having either a 48 amino acid deletion (*mre11ΔC48*) or a 136 amino acid deletion (*mre11-5*) are completely deficient in the introduction of meiotic double-stranded breaks at the initiation sites for meiotic homologous recombination [59, 60]. This deficiency is associated with the absence of a change in MNase hypersensitivity at double-stranded DNA break sites at the early stage of meiosis [60]. On the other hand, cells with these mutant Mre11 proteins are fully resistant to MMS treatment and proficient in telomere maintenance. This observation indicates that the putative C-terminal DNA binding sites are essential to the change in chromatin structure required for meiotic double-stranded cleavage, but these sites are not required for the mitotic functions of the Mre11 protein such as repair of double-stranded breaks and telomere maintenance.

5.4. Functional independence of the two domains of the Mre11 protein

The observations described above indicate that the Mre11 protein has two independent functional domains: the N-terminal nuclease domain, which is essential to mitotic functions and repair of meiotic double-stranded DNA-breaks, and the C-terminal DNA binding domain, which plays a role in the increased MNase hypersensitivity of chromatin DNA (chromatin transition) associated with meiotic double-stranded DNA breakage and the initiation of meiotic homologous recombination.

6. MEIOSIS-SPECIFIC CHROMATIN REMODELING AROUND THE *M26* MEIOTIC RECOMBINATION HOT SPOT AT THE *ade6* LOCUS OF *S. pombe*

6.1. Studies on meiotic recombination in *S. cerevisiae* and *S. pombe*

Although *S. cerevisiae* and *S. pombe* are called budding yeast and fission yeast, respectively, they are phylogenetically divergent [71]. Although the role of double-stranded breaks for the initiation of meiotic recombination has been well established in *S. cerevisiae*, meiosis-specific double-stranded breaks have not been detected in other organisms until their recent identification in *S. pombe* [72]. However, the relationship between meiotic double-stranded breakage and the initiation of meiotic homologous recombination has not yet been established in *S. pombe*. Thus, to test for a correlation between increased MNase hypersensitivity of chromatin DNA and the initiation of meiotic homologous recombination, we examined chromatin structure and its changes during meiosis in *S. pombe*. This analysis revealed another example of meiotic changes in chromatin structures that is essential to the initiation of meiotic homologous recombination.

6.2. Differences in the distribution of micrococcal nuclease hypersensitive sites depending on the presence or absence of the *ade6M26* meiotic recombination hot spot

As in *S. cerevisiae*, meiosis in *S. pombe* is synchronously induced by the deprivation of nitrogen sources from the culture medium. *S. pombe* diploid cells were grown in a nutritionally limited medium before the cells were transferred to a nitrogen-free medium and allowed to undergo meiosis and sporulation. In cells without the *M26* mutation, nucleosomes in the coding region of the *ade6* gene are regularly positioned with respect to the sequence (*i.e.*, phased). On the other hand, in cells with the *M26* recombination hot spot, the spacing of nucleosomes is disordered and an MNase-hypersensitive site is created at the site of the *M26* mutation [57]. This finding is not consistent with previous studies suggesting that chromatin structure in *M26* haploid vegetative cells is very similar to that in wild-type cells [73]. By examining the effects of growth conditions and ploidy, we found that the *M26* mutation-specific chromatin structure is induced under nutritionally limited conditions and only in diploid cells heterozygous at the mating locus [57] (K. Mizuno and K.O., unpublished observations). Thus, the *M26* mutation-specific chromatin structure, which includes an MNase-hypersensitive site at the *M26* mutation, represents extensive chromatin remodeling in diploid cells growing under conditions that are conducive to the induction of meiosis. Since this remodeling requires the complete heptamer sequence ATGACGT [57] and the presence of the active heterodimeric transcription factor Atf1-Pcr1 (K. O. and W. P. Wahls, unpublished observation), this remodeling is closely related to the meiotic recombination hot spot caused by the *M26* mutation.

The heptamer-dependent chromatin remodeling in meiosis is different from the Mre11 protein-dependent chromatin transition observed in *S. cerevisiae* for two reasons. Firstly, heptamer-dependent chromatin remodeling accompanies disordering of nucleosome positioning and results in a new MNase-hypersensitive site within a few hundred bases around the heptamer sequence, while the Mre11 protein-dependent transition occurs only at the *cis*-acting sequences that exhibit nuclease hypersensitivity. When the initiation sites were covered by phased nucleosomes (by removing an upstream transcription termination signal), the Mre11 protein-dependent chromatin transition was inhibited in *S. cerevisiae* [55].

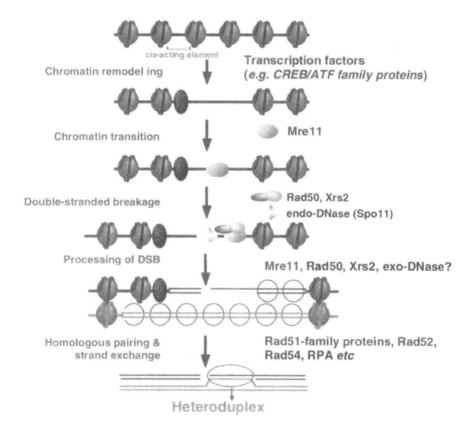

Fig. 2   A model of steps of double-stranded breakage for the initiation of meiotic recombination.   The model is based on our observations of the structural changes in chromatin structure around the *M26* recombination hotspot in *S. pombe* and those in *S. cerevisiae*.   The gene names are those of proteins of *S. cerevisiae*.

Secondly, heptamer-dependent remodeling around the *M26* mutation site is directly regulated by a transcription factor that recognizes a short consensus sequence, ATGACGT, while the chromatin transition at the *cis*-acting sequence for meiotic recombination is regulated by a homologous recombination-specific protein(s) (*i.e.*, the Mre11 protein) and proteins specifically involved in meiosis [58]. In addition, the *cis*-acting sequences in the latter case do not have a detectable consensus sequence.

6.3. Increased micrococcal nuclease hypersensitivity at the *M26* mutation site and the *ade6* promoter after the induction of meiosis

In addition to the chromatin remodeling around the *M26* meiotic recombination hot spot, the increase in MNase hypersensitivity at the *M26* hot spot and in the *ade6* promoter region, as in *S. cerevisiae* was observed [57]. The increase in MNase hypersensitivity in the promoter region was observed even in cells without the *M26* mutation. This increase in MNase hypersensitivity correlates with that observed in *S. cerevisiae*, suggesting that the chromatin transitions observed in the promoter of *ARG4* and other genes required for the initiation of meiotic homologous recombination is not unique to *S. cerevisiae*. Thus, in *S. pombe*, there are two stages of changes in chromatin structures around the initiation sites of meiotic homologous recombinations, *i.e.*, extensive chromatin remodeling to create MNase-hypersensitive sites under the regulation of a transcription factor that responds to starvation and diploidy, followed by site-specific chromatin transition (Fig. 2). In addition to tests whether the observed increase in MNase hypersensitivity at *M26* hot spot and the promoter region is dependent on *rad32*, an *S. pombe* gene encoding a homologue of *S. cerevisiae* Mre11 protein [74], biochemical and molecular structural analyses are awaited to understand the molecular basis of the local chromatin transition and the mechanistic role of the transition in the double-stranded break formation.

7. CONCLUDING REMARKS AND PERSPECTIVES

In eukaryotes, the initiation of homologous genetic recombination is regulated at the chromatin level by changes in its structure. At least two hierarchical levels of changes in chromatin structure are involved. The first level is extensive chromatin remodeling around the initiation sites for homologous recombination that is regulated by the ATF/CREB family transcription factor Atf1-Pcr1 under nutritional conditions and diploidy. This level of change was observed in *S. pombe* but not in *S. cerevisiae*. The involvement of the ATF/CREB family transcription factor in the induction of meiotic homologous recombination might reflect a role of homologous recombination in the adaptation of cells to environmental stresses. The second level is site-specific chromatin transition that is regulated by the Mre11 protein and other meiosis-specific regulatory proteins. This level of change was observed in both *S. cerevisiae* and *S. pombe*. It is likely that the higher order of chromatin structure prevents homologous recombination in higher eukaryotes. The initiation of homologous recombination in higher eukaryotes would involve the removal of such hierarchical structures to allow the access of proteins that initiate DNA recombination.

There is a need for methods that enable the targeted modification of genomic information, thereby allowing for effective gene therapy and the functional identification of isolated genes.

Homologous recombination fulfils these demands if one can activate homologous recombination in cells of higher eukaryotes, both mammals and plants. Agricultural plants and animals have been bred by crossing individuals with slightly different phenotypes acquired by spontaneous or artificial mutations. This is actually artificial acceleration of evolution using homologous recombination in a desired direction. However, this process is limited by the length of generation times. An understanding of the regulation and consequences of homologous recombination will contribute to the technological developments in these fields. In addition, based on this understanding, one can design a breeding strategy and precisely predict its consequences, thereby allowing an accurate risk-assessment.

ACKNOWLDEGEMENTS

The authors are grateful to Munenori Furuse (Genetic Dynamics Research Unit-Laboratory, RIKEN), Alain Nicolas (Curie Institute, Paris), Jürg Kohli (Bern University), Gerald R. Smith (Fred Hutchinson Cancer Research Center, Seattle) and Wayne P. Wahls (Vanderbilt University, Nashville) for their excellent contributions to this study. This study was supported by a grant from the "Biodesign Research Program" of RIKEN and partially by a grant for CREST from JST to T. S.

REFERENCES

1. A.J. Clark and A.D. Margulies, Proc. Natl. Acad. Sci. USA, 53 (1965) 451.
2. A.K.C. Wong, R. Pero, P.A. Ormonde, S.V. Tavtigian and P.L. Bartel, J. Biol. Chem., 272 (1997) 31941.
3. M. Kato, K. Yano, F. Matsuo, H. Saito, T. Katagiri, H. Kurumizaka, M. Yoshimoto, F. Kasumi, F. Akiyama, G. Sakamoto, H. Nagawa, Y. Nakamura and Y. Miki, J. Hum. Genet., 45 (2000) 133.
4. T. Horiuchi and Y. Fujimura, J. Bacteriol., 177 (1995) 783.
5. A. Kuzminov, Mol. Microbiol., 16 (1995) 373.
6. C.L. Doe, J. Dixon, F. Osman and M.C. Whitby, EMBO J., 19 (2000) 2751.
7. J.E. Haber, Trends Biochem. Sci., 24 (1999) 271.
8. C.M.J. Lovett, P.E. Love and R.E. Yasbin, J. Bacteriol., 171 (1989) 2318.
9. D. Dubnau, Microbiol. Rev., 55 (1991) 395.
10. K.N. Smith and A. Nicolas, Curr. Opin. Genet. Dev., 8 (1998) 200.
11. J.E. Haber, Cell, 89 (1997) 163.
12. G.S. Roeder, Proc. Natl. Acad. Sci. USA, 92 (1995) 10450.
13. N. Kleckner, Proc. Natl. Acad. Sci. USA, 93 (1996) 8167.
14. R.E. Esposito and S. Klapholz, in The molecular Biology of the yeast Saccharomyces. (J.N. Strathern, E.W. Jones and J.R. Broach, eds.), Cold Spring Harbor Laboratory, New York., pp. 211 (1981).
15. J.W. Szostak, T.L. Orr-Weaver, R.J. Rothstein and F.W. Stahl, Cell, 33 (1983) 25.
16. H. Sun, D. Treco, N.P. Schultes and J.W. Szostak, Nature (London), 338 (1989) 87.

17. L. Cao, E. Alani and N. Kleckner, Cell, 61 (1990) 1089.
18. F. Baudat and A. Nicolas, Proc. Natl. Acad. Sci. USA, 94 (1997) 5213.
19. T.H. Morgan and E. Cattell, J. Exp. Zool., 13 (1912) 79.
20. M. Meselson and J.J. Weigle, Proc. Natl. Acad. Sci. USA, 47 (1961) 857.
21. G. Kellenberger, M.L. Zichichi and J. Weigle, Proc. Natl. Acad. Sci. USA, 47 (1961) 869.
22. N. Anraku and J. Tomizawa, J. Mol. Biol., 12 (1965) 805.
23. R. Holliday, Genetic Res., Camb., 5 (1964) 282.
24. H. Potter and D. Dressler, Proc. Natl. Acad. Sci. USA, 73 (1976) 3000.
25. L. Bell and B. Byers, Proc. Natl. Acad. Sci. USA, 76 (1979) 3445.
26. K. McEntee, G.M. Weinstock and I.R. Lehman, Proc. Natl. Acad. Sci. USA, 76 (1979) 2615.
27. T. Shibata, C. DasGupta, R.P. Cunningham and C.M. Radding, Proc. Natl. Acad. Sci. USA, 76 (1979) 1638.
28. A. Shinohara, H. Ogawa and T. Ogawa, Cell, 69 (1992) 457.
29. D.K. Bishop, D. Park, L.Z. Xu and N. Kleckner, Cell, 69 (1992) 439.
30. A. Aboussekhra, R. Chanet, A. Adjiri and F. Fabre, Mol. Cell. Biol., 12 (1992) 3224.
31. G. Basile, M. Aker and R.K. Mortimer, Mol. Cell. Biol., 12 (1992) 3235.
32. A. Shinohara, H. Ogawa, Y. Matsuda, N. Ushio, K. Ikeo and T. Ogawa, Nat. Genet., 4 (1993) 239.
33. R. Cheng, T.I. Baker, C.E. Cords and R.J. Radloff, Mutat. Res., 294 (1993) 223.
34. P. Sung, Science, 265 (1994) 1241.
35. T. Ogawa, X. Yu, A. Shinohara and E.H. Egelman, Science, 259 (1993) 1896.
36. T.L. Orr-Weaver, J.W. Szostak and R.J. Rothstein, Proc. Natl. Acad. Sci. USA, 78 (1981) 6354.
37. A. Nicolas, D. Treco, N.P. Schultes and J.W. Szostak, Nature (London), 338 (1989) 35.
38. M. Lichten, C. Goyon, N.P. Schultes, D. Treco, J.W. Szostak, J.E. Haber and A. Nicolas, Proc. Natl. Acad. Sci. USA, 87 (1990) 7653.
39. T. Angel, B. Austin and D.G. Catchside, Aust. J. Biol. Sci., 23 (1970) 1229.
40. H. Gutz, Genetics, 69 (1971) 317.
41. P. Szankasi, W.D. Heyer, P. Schuchert and J. Kohli, J. Mol. Biol., 204 (1988) 917.
42. A.S. Ponticelli, E.P. Sena and G.R. Smith, Genetics, 119 (1988) 491.
43. P. Schuchert and J. Kohli, Genetics, 119 (1988) 507.
44. W.P. Wahls and G.R. Smith, Gene Develop., 8 (1994) 1693.
45. N. Kon, M.D. Krawchuk, B.G. Warren, G.R. Smith and W.P. Wahls, Proc. Natl. Acad. Sci. USA, 94 (1997) 13765.
46. M.E. Fox, T. Yamada, K. Ohta and G.R. Smith, Genetics, 156 (2000) 59.
47. T. Takeda, T. Toda, K.I. Kominami, A. Kohnosu, M. Yanagida and N. Jones, EMBO J., 14 (1995) 6193.
48. J. Kanoh, Y. Watanabe, M. Ohsugi, Y. Iino and M. Yamamoto, Genes Cells, 1 (1996) 391.
49. Y. Watanabe and M. Yamamoto, Mol. Cell. Biol., 16 (1996) 704.
50. H.L.K. Whitehouse, Nature (London), 211 (1966) 708.
51. B. de Massy and A. Nicolas, EMBO J., 12 (1993) 1459.
52. S. Keeney, C.N. Giroux and N. Kleckner, Cell, 88 (1997) 375.

53. A. Bergerat, B. deMassy, D. Gadelle, P.C. Varoutas, A. Nicolas and P. Forterre, Nature (London), 386 (1997) 414.
54. T.-C. Wu and M. Lichten, Science, 263 (1994) 515.
55. K. Ohta, T. Shibata and A. Nicolas, EMBO J., 13 (1994) 5754.
56. B. De Massy, V. Rocco and A. Nicolas, EMBO J., 14 (1995) 4589.
57. K.-i. Mizuno, Y. Emura, M. Baur, J. Kohli, K. Ohta and T. Shibata, Genes Dev., 11 (1997) 876.
58. K. Ohta, A. Nicolas, M. Furuse, A. Nabetani, H. Ogawa and T. Shibata, Proc. Natl. Acad. Sci. USA, 95 (1998) 646.
59. T. Usui, T. Ohta, H. Oshiumi, J. Tomizawa, H. Ogawa and T. Ogawa, Cell, 95 (1998) 705.
60. M. Furuse, Y. Nagase, H. Tsubouchi, K. Murakami-Murofushi, T. Shibata and K. Ohta, EMBO J., 17 (1998) 6412.
61. S. Moreau, J.R. Ferguson and L.S. Symington, Mol. Cell. Biol., 19 (1999) 556.
62. K.M. Trujillo, S.S. Yuan, E.Y. Lee and P. Sung, J. Biol. Chem., 273 (1998) 21447.
63. T.T. Paul and M. Gellert, Mol. Cell, 1 (1998) 969.
64. G.J. Sharples and D.R. Leach, Mol. Microbiol., 17 (1995) 1215.
65. J.C. Connelly, E.S. de Leau, E.A. Okely and D.R. Leach, J. Biol. Chem., 272 (1997) 19819.
66. J.C. Connelly and D.R. Leach, Genes Cells, 1 (1996) 285.
67. K. Nairz and F. Klein, Genes Dev., 11 (1997) 2272.
68. H. Sun, D. Treco and J.W. Szostak, Cell, 64 (1991) 1155.
69. C.I. Nugent, G. Bosco, L.O. Ross, S.K. Evans, A.P. Salinger, J.K. Moore, J.E. Haber and V. Lundblad, Curr. Biol., 8 (1998) 657.
70. S.J. Boulton and S.P. Jackson, EMBO J., 17 (1998) 1819.
71. G. Janbon, E.P. Rustchenko, S. Klug, S. Scherer and F. Sherman, Yeast, 13 (1997) 985.
72. D. Zenvirth and G. Simchen, Curr. Genet., 38 (2000) 33.
73. F. Bernardi, T. Koller and F. Thoma, Yeast, 7 (1991) 547.
74. M. Tavassoli, M. Shayeghi, A. Nasim and F.Z. Watts, Nucleic Acids Res., 23 (1995) 383.

PART III
PROTEIN FUNCTIONS

Molecular Anatomy of Cellular Systems
I. Endo et al., (editors)
© 2002 Elsevier Science B.V. All rights reserved.

Studies on photoreactive enzyme --nitrile hydratase--

Isao Endo[a] and Masafumi Odaka[b]

[a] Department of Bioproduction Science, Faculty of Agriculture, Utsunomiya University
350, Minecho, Utsunomiya-shi, Tochigi-ken 321-8505, Japan
[b] Bioengineering Laboratory, RIKEN
2-1, Hirosawa, Wako, Saitama 351-0198, Japan

Recent progress on the structure and function of a photoreactive enzyme, nitrile hydratase (NHase) was overviewed. NHase from *Rhodococcus* sp. N771 contains a non-heme iron center at the catalytic site, and its catalytic activity is regulated by nitrosylation and photo-induced denitrosylation of the iron center. Crystal structure analysis and mass spectrometry of nitrosylated NHase demonstrated the novel structure of the non-heme iron center having two post-translationally modified cysteine ligands, Cys-SOH and Cys-SO$_2$H. Studies on recombinant NHases proved that, at least, the Cys-SO$_2$H modification of αCys112 is responsible for the catalytic activity of NHase.

1. INTRODUCTION

The living organisms have used various metal elements in their substantial materials during the process of evolution. Obviously, many of these elements are toxic for the organisms if they are absorbed in excess. But appropriate quantities of them are essential for the organisms. Many researchers are studying why, where and how these metal elements are selected, combined, transported, stored and tuned to their specific functions. It is really wonderful achievement in life science that researchers have elucidated some physiological meanings of inorganic elements like Fe, Co, Cu, Mn, Mg, and so on for the living organisms. This studying field, it was born about a half century ago, is called the bioinorganic chemistry and developing rapidly today in the field of life science or biotechnology (1-4).

It is not the object of this article to survey neither advances in the bioinorganic chemistry nor more in narrow sense the metalloproteins that are incorporated metal elements in their principal parts of the compound. Obviously, activities of various metalloproteins cover energy, material and information transformation and expression of genes as well. Due to these facts, studies on chemical reaction scheme around the catalytic site of metalloprotein have been footlighted ultimately. The authors would like to focus specially on a photoreactive enzyme, nitrile hydratase, because the enzyme is important from the viewpoint of chemical engineering as well as bioinorganic chemistry.

Nitrile hydratase [NHase; EC.4.2.1.84] catalyzes the hydration of nitriles to the corresponding amides (5,6). The enzyme has been found in a number of microorganisms

160

including *Rhodococcus, Corrynebacterium, Nocardia, Arthrobacter, Pseudomonas* and *Klebsiella*, and some fungi, and classified into two groups with respect to the kinds of metal ions composing the catalytic center: Fe-type and Co-type. NHase originally found during the studies on microbial degradation of nitrile compounds by Yamada and his colleagues (7). Yamada and his colleagues and Nitto Chemical Industry Co. Ltd. (Presently, Mitsubishi Rayon Co. Ltd.) screened microorganisms that can use acrylonitrile as a sole carbon source as well as nitrogen one and obtained various bacterial strains exhibiting the NHase activity (5,6). They adopted initially Fe-type NHase for the industrial production of acrylamide from acrylonitrile, and later, did Co-type NHase because of the stability and high yield. Figure 1 shows two industrial production of acrylamide. (a) shows the *classic process* which applies Cu catalyst. Because of low efficiency in nitrile hydration reaction and exceeding production of by-products, acrylamide produced must be purified and Cu catalyst, which is harmful toward the environment, should be recycled. They have to construct a very complex process to prevent eluviation of the catalyst from the plant to environment. On the contrary, the process (b) is very simple because of the high-yield acrylamide production of the biocatalyst, Co-type NHase. Due to application of the biocatalyst, this process is very mild toward the environment. Presently, one-third of industrial acrylamide production in the world was preformed by using this process. This manufacture is the first successful example of a bioconversion process for the production of commodity chemicals.

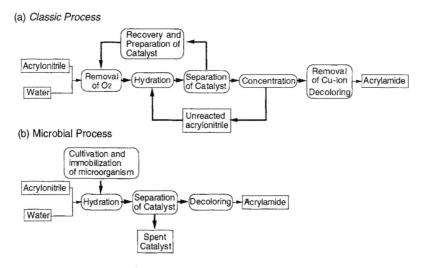

Figure 1. Comparative Flowsheet for (a) the *classic process* using Cu catalyst and (b) microbial one.

In the course of screening of the biocatalyst, Nitto Chemical Industry Co. Ltd. found curious strains whose NHase activity greatly depends on the culture condition: NHase activity of the strains was very high when they were cultured in glass flask, but it dramatically decreased when did in stainless jar fermenter. Rightly, Nitto Chemical Industry

Co. Ltd. gave up the application of the strain for the acrylamide production. One of the authors, however, had been attracted by this unique characteristics of the strain and borrowed from the company one of the strain, *Rhodococcus* sp. N771, for academic use. Instantly, we have found that NHase from *Rhodococcus* sp. N771 shows a unique photoreactivity; the enzyme is inactivated by aerobic incubation in the dark whereas it is activated by photo-irradiation (8,9).

Here in this chapter, we describe studies on the structure and function of the photoreactive nitrile hydratase.

2. PHOTOREACTIVITY OF FE-TYPE NHASE

NHase is composed of two kinds of subunits (α and β) with a basic stoichiometry of $\alpha_i\beta_i M_i$ (M, Fe^{3+} or Co^{3+}). Their primary sequences are well conserved among all known NHases while there is no apparent homology between the two subunits (6). Electron spin resonance (ESR) (10,11) and X-ray absorption spectroscopy (11,12) have proved that Fe- and Co-type NHases have non-heme iron(III) and non-corrin cobalt(III) atoms at the catalytic center, respectively, and that the ligand fields of both types are similar. Photo-responsible activation of NHase *in vivo* have been reported only in three strains, *Rhodococcus* sp. N771 (8,9), N774 (13) and R312 (12). These NHases are Fe-type one and provably the same protein because the nucleotide sequences of their genes are identical (14-16).

In the early stage of physicochemical studies on the *Rhodococcus* sp. N771 NHase, UV-Vis (9,17), ESR (18), and Mössbauer (19) spectroscopic measurements suggested the involvement of the non-heme iron center in the photo-reactivation process. Then, one of our co-workers, Noguchi, examined a light-induced difference Fourier-transform infrared (FT-IR) spectrum of the NHase (20). In addition to a number of bands showing small conformational changes in the frequency region lower than 1700 cm^{-1}, large bands at about 1850 cm^{-1}, where biochemical compounds exhibit no signal in normal, were observed. Nitric oxide (NO) was considered as a candidate for the compound causing these signals. The hypothesis was confirmed by ^{15}N-substitution. This is the first experimental evidence of photo-reactivation involving NO in the biological knowledge (20). The reversible conversion

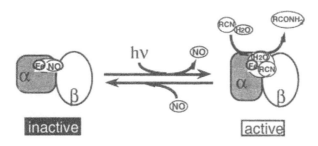

Figure 2. Scheme of photoreactivity of NHase from *Rhodococcus* sp. N771.

of the photo-reactivated NHase into the inactive one by nitrosylation was confirmed by FT-IR, ESR, UV-Vis absorption, and resonance Raman measurements (21, 22). The formation of the 1:1-stoichiometry of nitrosyl-iron complex was confirmed both in the *Rhodococcus* sp. R312 NHase and in a synthetic model compound by X-ray absorption spectroscopy (23). The mechanism of photo-reactivation and dark-inactivation is drawn schematically in Figure 2.

Rhodococcus sp. N771 should produce NO endogenously. NHase is nitrosylated to form a very stable inactive complex in the dark, and the NO ligand is photo-dissociated and released to the solvent upon photoactivation. The Fe-type NHase from *Comamonas testosteroni* NI1, which does not respond to light *in vivo*, also exhibits the photo-reactivity by being exposed into exogenous NO (24). In contrast, Co-type NHase has never shown to exhibit any photoreactivity by nitrosylation even *in vitro* (M. Tsujimura, M. Nojiri and M. Odaka, unpublished results). Probably, the photo-reactivity ruled by NO is a common characteristic of Fe-type NHases.

A biologically important question arises: where NO molecules come from in the microorganism? Sari *et al.* (25) reported the existence of a nitric oxide synthase, catalyzing conversion of L-arginine into L-citrulline and NO, in the crude extract of *Rhodococcus* sp. R312. However, we found that the crude extract from *Rhodococcus* sp. N771 showed no activity to convert [^{14}C]-L-arginine into [^{14}C]-L-citrulline (M. Odaka, H. Hori, T. Nishino, M. Yohda and I. Endo, unpublished results). Although the NO producing system in this microorganism has not been determined, it is true that NO regulates the photoreactivity of Fe-type NHase.

3. STRUCTURE DETERMINATION OF THE NOVEL NON-HEME IRON CENTER OF NHASE

3.1. Post-translational modification of αCys112 to cysteine-sulfinic acid

The non-heme iron center is not only the photoreaction center but also the catalytic site of NHase. Then, how is its structure? The structure of the iron center had been extensively studied by various spectroscopic measurements including ESR (10), electron-nuclear double resonance (ENDOR) (26,27), resonance Raman (28,29) and X-ray absorption spectroscopy (30, 31). The distorted six-coordinate structure with two cysteine thiolates and three histidine immidazoles and a solvent molecule was suggested. We found the nitrosyl iron center of *Rhodococcus* sp. N771 is stable even in the presence of 6 M urea (32), and then isolated the tryptic peptide fragment containing the nitrosyl iron center (33). The protein sequencing

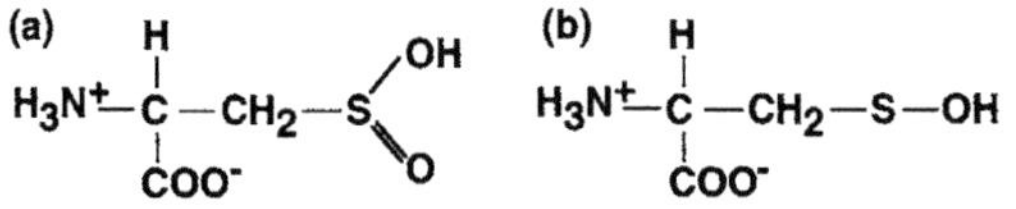

Figure 3. Structure of (a) cysteine-sulfinic acid and (b) -sulfenic acid.

analysis has demonstrated that the iron center exists in the region from αAsn105 to αLys128. The region contains three cysteine residues (αCys109, αCys112 and αCys114) but did no histidine residue. Interestingly, mass spectrometry of the corresponding peptide isolated from the photoactivated NHase revealed that αCys112 is post-translationally modified to cysteine-sulfinic acid (Cys-SO$_2$H, Fig. 3a) (33). So long as we know, Fe-type NHases is the first example of the enzyme having the post-translationally modified cysteine ligand in a native protein.

3.2. Crystal structure analysis of Fe-type NHase I

Recently, the crystal structure of the Fe-type NHase in the inactive state at a resolution of 1.7 Å (34, 35) and that in the photo-reactivated state at 2.65 Å (36) were reported. Overall structure and the ligand atoms were identical in both structures. The non-heme iron center is located at the interface of the two subunits. The sulfur atoms from αCys109, αCys112 and αCys114 and two amide nitrogen atoms of αSer113 and αCys114 are coordinated to the iron both in the nitrosylated and photoactivated states. The sixth coordination site, which is thought to be the substrate-binding site, is occupied by an NO molecule in the crystal structure of the nitrosylated state while this site is vacant in that of the photoactivated one.

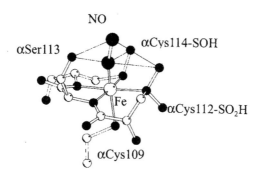

Figure 4.   Structure of the non-heme iron active center of NHase from *Rhodococcus* sp. N771 in the nitrosylated state. Atom pairs within hydrogen bonding distance (<3.4 Å) are linked with thin lines. The iron-ligand interactions are represented with gray sticks.

The structure around the non-heme iron center in the inactive state was shown in Figure 4. In addition to the modification of αCys112-SO$_2$H (32), the crystal structure demonstrated that αCys114 was post-translationally modified into a cysteine sulfenic acid (Cys-SOH, Fig. 3b) in the nitrosylated NHase (34). The modification of αCys114 was verified by Fourier-transform ion cyclotron resonance mass spectrometry against the iron center peptide prepared from the nitrosylated enzyme (34). Three oxygen atoms, Oδ1 of αCys112-SO$_2$H, Oδ of αCys114-SOH and Oγ of αSer113, protrude from the plane containing the iron atom and hold an endogenous NO molecule on the top of the iron center like claws (Fig. 4). Therefore, we termed this unprecedented structure as "claw setting". This structure likely provides a suitable explanation on the extraordinary stability of the nitrosyl ferric-iron complex of NHase, which exhibits persistent stability under aerobic condition for over a year in the dark.

To date, the Fe-type NHase from *Rhodococcus* sp. N-771 is the only metalloenzyme that has the post-translationally modified cysteine ligands in its metal site in a native protein. However, the metal-binding motif (consensus sequence: Cys1-Xxx-Leu-Cys2-Ser-Cys3 (Xxx = Ser(Fe-type) / Thr(Co-type)), is completely conserved among all known NHases (14-16, 37-40) and even in the homologous enzyme, thiocyanate hydrolase (41) (Fig. 5). Thus, certain enzymes with the sequence motif are likely to possess similar post-translational modifications, and constitute the NHase family.

3.3. Crystal structure analysis of Fe-type NHase II

The crystal structures introduced important questions against the structure around the iron center. Firstly, no modification was observed in the crystal structure in the photoactivated state. Secondly, the vacancy of the sixth ligand site is inconsistent with the previous ESR and ENDOR studies on the photo-activated NHase, from which the sixth ligand-binding site is

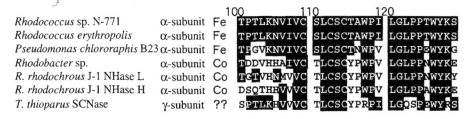

Figure 5. Amino acid sequence alignments near the metal binding motif of various NHases and thiocyanate hydrolase. The residue numbers are in accordance with those in Fe-type NHase of *Rhodococcus* sp. N771. Black backgrounds represent the amino acid residues conserved with Fe-type NHase of *Rhodococcus* sp. N771.

demonstrated to be occupied a water or hydroxyl group. Thirdly, n-butyric acid, which is widely used as a stabilizing agent of Fe-type NHases during purification, storage and most experiments that have been reported, could not be observed in the crystal structure of the photo-activated NHase, despite that the crystal was grown in the presence of enough concentration of n-butyric acid. These discrepancies probably due to the low resolution and poor statistics of the crystal used for crystal structure determination of the photo-activated NHase. Obviously, the high-resolution structure analysis of the NHase in the photo-activated state should be indispensable. Thus, we crystallized the photo-activated NHase with a new crystal form and determined the crystal structure of the photo-activated NHase at 1.5 Å resolution. We also identified the sixth ligand atom and the localization of n-butyric acid. We found the post-translational modifications of the cysteine ligands in the photo-activated NHase for the first time. Interestingly, Cys114 existed as a Cys-SO$_2$H, suggesting that Cys-114-SOH was further oxidized during crystallization. These studies will be published elsewhere.

4. RECENT STUDIES ON RECOMBINANT NHASES

4.1. Expression of recombinant NHase in *E. coli*

We have determined the structure of the NHase operon in *Rhodococcus* sp. N771 (Fig. 6) [21]. The operon consists of two open reading frames (named as *nhr1* and *nhr2*), amidase (*ami*), NHase α-subunit (*nha1*), NHase β-subunit (*nha2*) and an open reading frame (*nha3*). The nhr1 and nhr2 region showed strong similarities with the regulators of the *Rhodococcus rhodochrous* J1 L-NHase (42). nha3 encodes a polypeptide of 396 amino acids homologous to P47k, which is encoded by a gene in the NHase operon of *Pseudomonas chlororaphis* B23 and responsible for functional expression of NHase (38). We constructed NHase expression plasmids, pRCN102 and pRCN104, using T7 expression system (Fig. 6) (16). The former contains only the α– and the β–subunit genes and the latter does *nha3* in addition to the subunit genes. The *E. coli* transformant harboring pRCN102 expressed NHase proteins only as the inclusion bodies, while that harboring pRCN104 showed significant NHase activity when it was cultivated at 30 °C. After transforming *E coli* harboring pRCN104 with pHSGβ, the expression plasmid for the NHase β–subunit (Fig. 6), we succeeded to express functional NHase up to 25 % of total soluble protein in the *E. coli* transformant by cultivating the transformant at less than 30 °C (typically at 27 °C). The recombinant NHase showed NHase activity comparable to the native one and reversibly inactivated by exogenous NO. Mass spectrometry of the iron center peptide revealed that αCys112 of the recombinant NHase was post-translationally modified into Cys-SO$_2$H.

These results indicate clearly that the protein encoded by *nha3* is essential for the functional expression of Fe-type NHase. We named this protein as NHase activator. NHase activator may be required for the incorporation of an iron ion, and/or involved in the modification of αCys112 and/or αCys114. The study on the role of the activator is currently underway in our laboratory.

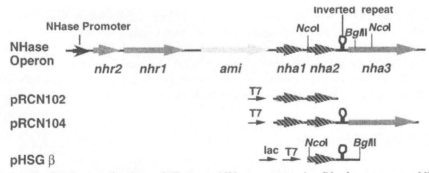

Figure 6. Gene organization of Fe-type NHase operon in *Rhodococcus* sp. N771 and schematic representation of the plasmids used.

4.2. Role of the post-translationally modified cysteine residues in recombinant NHases

Recently, we have succeeded in reconstituting the $\alpha\beta$ complex *in vitro* from the recombinant α- and β-subunits, which contain no modification, in the presence of a ferric ion under anaerobic conditions (43). The reconstitution mixture did not contain the NHase activator. The obtained $\alpha\beta$-complex exhibited no enzymatic activity at the time of reconstitution, but gradually acquired the activity during aerobic incubation. Mass spectrometry of the iron center peptide isolated from the $\alpha\beta$-complex before and after aerobic incubation showed that αCys112 was modified to Cys-SO_2H during aerobic incubation. The modification of αCys114 to Cys-SOH could not be confirmed because of its instability. These results indicate that the modification of αCys112 to Cys-SO_2H is essential for the catalytic activity of Fe-type NHase.

Co-substituted Fe-type NHase (NHase$_{(Fe \to Co)}$) was obtained by cultivation of the *E. coli* transformant harboring the NHase expression vector lacking the NHase activator gene on Luria-Bertani medium supplemented with cobalt(II) chloride (44). The purified NHase$_{(Fe \to Co)}$ exhibited rather weak NHase activity, but this activity was enhanced by a factor of 2.5 after oxidation with potassium hexacyanoferrate. The UV-Vis spectral changes of NHase$_{(Fe \to Co)}$ induced by the oxidation suggest that the Co center was converted to the low-spin Co^{3+} state. The relative amount of NHase$_{(Fe \to Co)}$ having the Cys-SO_2H modification of αCys112 was greatly increased after the oxidation. Thus, it was concluded that αCys112-SO_2H stabilization as well as the low-spin Co^{3+} state are essential for the activity of NHase$_{(Fe \to Co)}$.

5. CONCLUSION

In this chapter, we have shown recent progresses on a photoreactive NHase. The photoreactivity of Fe-type NHase is controlled by binding and release of NO. Crystal structure analysis and mass spectrometry of the nitrosylated NHase have unveiled the unprecedented structure of the non-heme iron center having two post-translationally modified cysteine ligands, Cys-SOH and Cys-SO_2H. Studies on recombinant NHases demonstrated that, at least, the Cys-SO_2H modification of αCys112 is responsible for the catalytic activity of NHase. The detailed role of each modification is a subject for further investigation. The conservation of the modifications in members of the NHase family has not been confirmed yet. However, the high homology of the amino acid sequences near the metal binding motif suggests that the structure of the metal center including the modifications is common in members of the NHase family. Studies on the enzymes of the NHase family will provide insight into the biological functions of the Cys-SOH and Cys-SO_2H modifications in proteins.

6. ACKNOWLEDGEMENTS

We sincerely thank Drs. M. Tsujimura, H. Nakayama, M. Yohda, Y. Kawano, N. Dohmae, M. Nakasako, S. Nagashima, K. Takio, K. Takio, N. Kamiya, T. Nagamune, M. Kobayashi, T. Noguchi, M. Hoshino and many students for their energetic contribution to the studies. We thank Dr. M. Maeda for valuable discussion and suggestions. We also extended our thanks to the Biodesign Research Program, the Essential Chemical Reactions Research Program, and the Bioarchitect Research Program in RIKEN.

REFERENCES

1. J. J. R. Fraústo da Silva and R. J. P. Williams, in The Biological Chemistry of the Elements – The Inorganic Chemistry of Life, Clarendon press, Oxford (1991).
2. I. Bertini, H. B. Gray, S. J. Lippard and J. S. Valentine, in Bioinorganic Chemistry, Universal Science Books, Mill Valley, California (1991).
3. S. J. Lippard, and J. M. Berg, in Principles of Bioinorganic Chemistry, Universal Science Books, Mill Valley, California (1994).
4. I. Bertini, A. Sigel and H. Sigel, Handbook of Metalloproteins, Marcel Dekker, Basel, New York, (2001).
5. M. Kobayashi and S. Shimizu, Nature Biotechnol., 16 (1998) 733.
6. M. Kobayashi, T. Nagasawa, H. Yamada, Trends Biotechnol., 10 (1992) 402.
7. Y. Asano, Y. Tani and H. Yamada, Agricultural Biol. Chem., 44 (1980) 2251.
8. T. Nagamune, H. Kurata, M. Hirata, J. Honda, A. Hirata, I. Endo, Photochem. Photobiol., 51 (1990) 87.
9. T. Nagamune, H. Kurata, M. Hirata, J. Honda, H. Koike, M. Ikeuchi, Y. Inoue, A. Hirata, I. Endo, Biochem. Biophys. Res. Commun., 168 (1990) 437.
10. Y. Sugiura, J. Kuwahara, T. Nagasawa, H. Yamada. J. Am. Chem. Soc., 109 (1987) 5848.
11. R.C. Scarrow, B.A. Brennan, J.G. Cummings, H. Jin, Duong D., J.T. Kindt, M. J. Nelson, Biochemistry, 35 (1996) 10078.
12. B. A. Brennan, G. Alms, M. J. Nelson, L. T. Durney, R. C. Scarrow, J., Am. Chem. Soc., 118 (1996) 9194.
13. T. Nakajima, K. Takeuchi, H. Yamada, Chem. Lett., 9 (1987) 1767.
14. O. Ikehata, M. Nishiyama, S. Horinouchi, T. Beppu, Eur. J. Biochem., 181 (1989) 563.
15. J.F. Mayaux, E. Cerebelaud, F. Soubrier, D. Faucher, D. Petre, J. Bacteriol., 172 (1990) 6764.
16. M. Nojiri, M. Yohda, M. Odaka, Y. Matsushita, M. Tsujimura, T. Yoshida, N. Dohmae, K. Takio, I. Endo, J. Biochem., 125 (1999) 696.
17. J. Honda, H. Kandori, T. Okada, T. Nagamune, Y. Shichida, H. Sasabe, I. Endo Biochemistry, 33 (1994) 3577.
18. J. Honda, T. Nagamune, Y. Teratani, A. Hirata, H. Sasabe, I. Endo, Ann. N.Y. Acad. Sci., 672 (1992) 29.
19. J. Honda, Y. Teratani, Y. Kobayashi, T. Nagamune, H. Sasabe, A. Hirata, F. Ambe, I. Endo, FEBS Lett., 301 (1992) 177.
20. T. Noguchi, J. Honda, T. Nagamune, H. Sasabe, Y. Inoue, I. Endo, FEBS Lett., 358 (1995) 9.
21. M. Odaka, K. Fujii, M. Hoshino, T. Noguchi, M. Tsujimura, S. Nagashima, M. Yohda, T. Nagamune, Y. Inoue, I. Endo, J. Am. Chem. Soc., 119 (1997) 3785.
22. T. Noguchi, M. Hoshino, M. Tsujimura, M. Odaka, Y. Inoue, I. Endo, Biochemistry, 35 (1996) 16777.
23. R.C. Scarrow, B.S. Strickler, J.J. Ellison, S.C. Shoner, J.A. Kobacs, J.G. Cummings, M.J. Nelson, J. Am. Chem. Soc., 120 (1998) 9237.
24. D. Bonnet, I. Artaud, C. Moali, D. Pétré, D. Mansuy, FEBS Lett., 409 (1997) 216.

25. M.-A. Sari, C. Moali, J.-L. Boucher, M. Jaouen, D. Mansuy, Biochem. Biophys. Res. Commun., 250 (1998) 364.

26. H. Jin, I.M. Turner, Jr., M.J. Nelson, R.J. Gurbiel, P.E. Doan, B.M. Hoffman, J. Am. Chem. Soc. 115 (1993) 5290.

27. P.E. Doan, M.J. Nelson, H. Jin, B.M. Hoffman, J. Am. Chem. Soc. 118 (1996) 7014.

28. H. Jin, I. M. J. Turner, M. J. Nelson, R. J. Gurbiel, P. E. Doan, B. M. Hoffman, J. Am. Chem. Soc., 115 (1993) 5290.

29. B. A. Brennan, J. G. Cummings, D. B. Chase, I. M. Turner, Jr., M. J. Nelson, Biochemistry, 35 (1996) 10068.

30. M. J. Nelson, H. Jin, I. M. J. Turner, G. Grove, R. C. Scarrow, B. A. Brennan. and L. Que, Jr., J. Am. Chem. Soc., 113 (1991) 7072.

31. R. C. Scarrow, B. A. Brennan, J. G. Cummings, H. Jin, D. J. Duong, J. T. Kindt, M. J. Nelson, Biochemistry 35 (1996) 10078.

32. M. Tsujimura, M. Odaka, S. Nagashima, M. Yohda and I. Endo, J. Biochem. (Tokyo), 119 (1996) 407.

33. M. Tsujimura, N. Dohmae, M. Odaka, M. Chijimatsu, K. Takio, M. Yohda, M. Hoshino, S. Nagashima and I. Endo, J. Biol. Chem., 272 (1997) 29454.

34. S. Nagashima, M. Nakasako, N. Dohmae, M. Tsujimura, K. Takio, M. Odaka, M. Yohda, N. Kamiya and I. Endo, Nature Struct. Biol., 5 (1998) 347.

35. M. Nakasako, M. Odaka, M. Yohda, N. Dohmae, K. Takio., N. Kamiya and I. Endo, Biochemistry, 38 (1999) 9887.

36. W. Huang, J. Jia, J. Cummings, M. Nelson, G. Schneider and Y. Lindqvist, Structure 5 (1997) 691.

37. M.S. Payne, S. Wu, R.D. Fallon, G. Tudor, B. Stiglitz, I.M. Turner, Jr. and M.J. Nelson, Biochemistry, 36 (1997) 5447.

38. M. Nishiyama, S. Horinouchi, M. Kobayashi, T. Nagasawa, H. Yamada and T. Beppu, J. Bacteriol., 173 (1991) 2465.

39. M. Kobayashi, M. Nishiyama, T. Nagasawa, S. Horinouchi, T. Beppu, H. Yamada, Biohcim. Biophys. Acta, 1129 (1991) 23.

40. B. Proou, K. Yamada, H. Morimoto, Patent JP 1994303971-A2.

41. Y. Katayama, Y. Matsushita, M. Kaneko, M. Kondo, Mizuno T. and H. Nyunoya, J. Bacteriol., 180 (1998) 2583.

42. H. Komeda, M. Kobayashi and S. Shimizu, J. Biol. Chem., 271 (1996) 15796.

43. T. Murakami, M. Nojiri, H. Nakayama, M. Odaka, M. Yohda, N. Dohmae, K. Takio, T. Nagamune and I. Endo, Protein Science, 9 (2000) 1024.

44. M. Nojiri, H. Nakayama, M. Odaka, M. Yohda, K. Takio and I. Endo, FEBS Lett., 465 (2000) 173.

Molecular Anatomy of Cellular Systems
I. Endo et al., (editors)

Structural and functional analyses of proteins involved in translation, DNA recombination, chromosome architecture, and signal transduction

Hitoshi Kurumizaka[a, b] and Shigeyuki Yokoyama[a, b, c]

[a] Cellular Singnaling Laboratory, RIKEN Harima Institute at SPring8
1-1-1 Kohto, Mikazuki-cho, Sayo, Hyogo 679-5143, Japan
[b] RIKEN Genomic Sciences Center
1-7-22 Suehiro-cho, Tsurumi, Yokohama 230-0045, Japan
[c] Department of Biophysics and Biochemistry, Graduate School of Science, University of Tokyo, 7-3-1 Hongo, Bunkyo-ku, Tokyo 113-0033, Japan

Structural biology, in which the mechanisms of biological systems are elucidated by the three-dimensional structures of biomacromolecules, such as proteins and nucleic acids, is rapidly progressing. An aim of our research is to establish a novel strategy for understanding the functions of biomacromolecules through structural biology as well as biochemistry, molecular biology, and cell biology. In this article, we discuss our current progress in structural biology toward understanding the mechanisms of translation, DNA recombination, centromere architecture, and signal transduction.

1. TRANSLATION

Genetic information is transcribed from DNA to RNA and is subsequently translated into proteins. In this gene expression process, the proteins that interact with RNA play crucial roles. In this section, we discuss the structural details of how the proteins, aminoacyl-tRNA synthetases (aaRSs), and an mRNA-binding protein regulate the translational process.

1.1. Structural basis for double-sieve discrimination between L-valine and L-isoleucine by aminoacyl-tRNA synthetases

The aaRSs aminoacylate the cognate tRNA with the cognate amino acid, which will be incorporated into the nascent polypeptide chain on the ribosome. High fidelity in protein synthesis is vitally important for all biological systems. The specificity between the amino acid and the codon (the genetic code) critically depends on the aminoacylation reactions catalyzed by the aaRSs (1). In this reaction, the aaRSs esterify cognate amino acids with their specific tRNAs, which decode trinucleotide sequences (anticodons) into amino acids corresponding strictly to the codons. The amino acid and adenosine triphosphate (ATP) form an aminoacyl adenylate as an active intermediate, and the aminoacyl moiety is then transferred to the 3' terminal adenosine of the tRNA.

Discrimination between L-isoleucine and L-valine is one of the most difficult recognitions to achieve, because they differ by only one methylene group in their aliphatic side chains. Actually, IleRS has an additional editing activity that hydrolyzes both valyl-adenosine monophosphate (Val-AMP) and Val-tRNA^{Ile} in a tRNA^{Ile}-dependent manner (2). In order to explain the strict L-isoleucine and L-valine discrimination, which is thermodynamically impossible through an ordinary one-step recognition, Fersht proposed the "double-sieve" (two-step substrate selection) model for the mechanism of the amino-acid selection by IleRS (2): amino acids larger than L-isoleucine are excluded by the aminoacylation site, serving as the "first, coarse sieve," and smaller ones, such as L-valine, are strictly eliminated by the "second, fine sieve" of the putative "editing site".

To understand the molecular mechanisms of double-sieve selection, we determined the crystal structure of *Thermus thermophilus* IleRS (1045 amino acid residues, 120 kD) and those of the complexes of IleRS with L-isoleucine and L-valine at resolutions of 2.5, 2.8, and 2.8 Å, respectively (3). IleRS is a member of the class I synthetases, which are characterized by an ATP-binding domain constructed with a Rossmann fold. *Thermus thermophilus* IleRS is a thick, L-shaped molecule with an approximate size of 100 Å by 80 Å by 45 Å (Fig.1A). The IleRS structure exhibits the Rossmann-fold domain at the center, β-rich intervening domains at the top, and an α-rich cylindrical domain at the bottom. The Rossmann-fold domain has a central deep catalytic cleft, and one L-isoleucine molecule is bound at the bottom of this catalytic cleft. The hydrophobic side chain of L-isoleucine is recognized by a pocket consisting of Pro46, Trp518, and Trp558 through van der Waals interactions. Interestingly, the L-leucine side chain does not fit into this pocket, because of the steric hindrance of one of the terminal methyl groups. In contrast, an L-valine molecule is actually bound to the same site in the L-valine·IleRS complex structure. These results agree with the concept of the first, coarse sieve in the double-sieve mechanism of editing.

For ValRS, a model analogous to the "double-sieve" model for the IleRS editing has been proposed (2): L-threonine and L-valine are recognized by the shape in the first step, and L-threonine is discriminated from L-valine by the presence of the hydroxyl group in the second step. In our crystal structure of the complex of *Thermus thermophilus* ValRS, tRNA^{Val}, and an analog of the Val-adenylate intermediate (Fig. 1B)(4), the analog is bound in a pocket, where Pro41 allows accommodation of the Val and Thr moieties, but precludes the Ile moiety (the first sieve), on the aminoacylation domain. The editing domain, which hydrolyzes incorrectly synthesized Thr-tRNA^{Val}, is bound to the 3' adenosine of tRNA^{Val}. A contiguous pocket was found to accommodate the Thr moiety, but not the Val moiety (the second sieve). These structural studies with IleRS and ValRS have elucidated how the aaRS·tRNA complex achieves the sophisticated "double-sieve" discrimination of L-isoleucine and L-valine from the other amino acids during translation.

1.2. Anticodon recognition by the discriminating glutamyl-tRNA synthetase

Given the importance of accurate translation, each organism would be expected to have at least twenty aminoacyl-tRNA synthetases. However, in fact, most bacteria, archaea, chloroplasts, and mitochondria lack the aminoacyl-tRNA synthetase specific to glutamine (GlnRS)(5-8). In these nineteen-synthetase systems, the aminoacyl-tRNA synthetase specific to glutamic acid (GluRS) aminoacylates both tRNA^{Glu} and tRNA^{Gln} with glutamic acid ("non-discriminating" GluRS), and the "misacylated" product, Glu-tRNA^{Gln}, is converted to Gln-tRNA^{Gln} by a transamidation enzyme. The twenty-synthetase systems changed their GluRS to be able to discriminate against tRNA^{Gln} and, therefore, to specialize in tRNA^{Glu} ("discriminating" GluRS)(9-11).

A

B

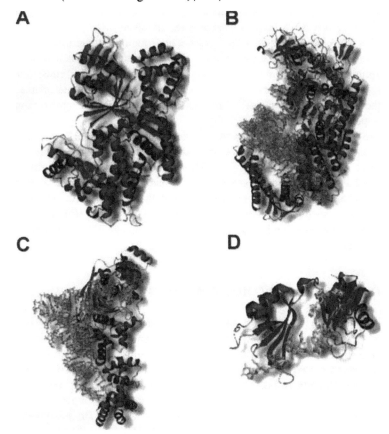

C

D

Figure 1 Crystal structures of the IleRS (A), ValRS•tRNA (B), GluRS•tRNA (C), and Sxl•mRNA. The α-helices, β-sheets, and RNA are shown in red, green, and yellow, respectively.

To elucidate the structural basis of tRNA recognition and discrimination by the "discriminating" GluRS, the crystal structures of the GluRS itself and the GluRS·tRNA^{Glu} complex were determined (Fig.1C)(12).   In the complex, the GluRS recognizes the tRNA^{Glu} anticodon bases *via* two α-helical domains, maintaining the base stacking.   The discrimination between the Glu and Gln anticodons (^{34}YUC^{36} and ^{34}YUG^{36}, respectively) is achieved by a single arginine residue (Arg358).   The mutation of Arg358 to Gln resulted in a GluRS that does not discriminate between the Glu and Gln anticodons (12).   This change mimics the reverse course of GluRS evolution from anticodon "non-discriminating" to "discriminating".

1.3. Recognition of the *tra* mRNA precursor by the Sex-lethal protein

In eukaryotic cells, mRNA precursors (pre-mRNA) undergo a series of post-transcriptional events, such as processing, splicing, translocation to the cytoplasm, translation, and degradation, in which numerous RNA-binding proteins are involved (13). The *Sex-lethal* (*Sxl*) gene product is an RNA-binding protein that plays a key role in sex determination and dosage compensation in *Drosophila melanogaster*(14). The Sxl protein induces the female-specific alternative splicing of the *transformer* (*tra*) pre-mRNA (15, 16). Sxl binds tightly to a uridine-rich polypyrimidine tract at the non-sex-specific 3' splice site in one of the *tra* introns, preventing the general splicing factor U2AF from binding to this site, and forcing it to bind to the female-specific 3' splice site (17).

The crystal structure of the complex between the two tandemly arranged RBDs of the Sxl protein and a 12-nucleotide single-stranded RNA derived from the *tra* polypyrimidine tract has been determined at 2.6Å resolution (Fig. 1D)(18). In the crystal structure, the two RNA-binding domains have their β-sheet platforms facing each other to form a V-shaped cleft. The RNA is characteristically extended and bound in this cleft, where the UGUUUUUU sequence is specifically recognized by the protein. This structure offers new insight into how a protein binds specifically to a cognate RNA without any intramolecular base-pairing.

2. DNA RECOMBINATION AND CHROMOSOME ARCHITECTURE

Double-strand breaks (DSBs) in chromosomal DNA frequently occur when cells are exposed to various DNA-damaging agents, including ionizing radiation, crosslinking reagents, and oxidative stress. DSBs are also observed when cells undergo a meiotic cell division cycle. In eukaryotes, two major pathways that repair lethal chromosomal DSBs have been identified: homologous recombination and nonhomologous DNA end joining. The homologous recombination pathway is important for accurate DNA repair, which is indispensable for chromosomal maintenance. In this section, we focus on the machinery for homologous pairing, which is a key step in the homologous recombination pathway .

On the other hand, the centromere is an essential region of the chromosome for its segregation during cell division, and has a special chromatin structure involving α-satellite DNA repeats. To understand the architecture of human chromosomes, we are currently trying to reconstitute the centromere with its specific proteins, a histone H3 variant CENP-A and CENP-B, and the core histones, H2A, H2B, and H4. In this section, we introduce the structure of a DNA-binding region of CENP-B.

2.1. The human Rad51·Rad52 interaction

Homologous pairing is a process in which a single-stranded DNA (ssDNA) tail derived from a DSB site invades the homologous double-stranded DNA (dsDNA) to form a heteroduplex. In bacteria, RecA, consisting of 352 amino acid residues, promotes the homologous pairing step (19, 20). A eukaryotic RecA homologue, Rad51, which is highly conserved from yeast to human (21, 22), has been discovered as a gene product in the *RAD52* epistasis group. A mutation in any of the *RAD52* group of genes (*RAD50*, *RAD51*, *RAD52*, *RAD54*, *RAD55*, *RAD57*, *RAD59*, *MRE11*, and *XRS2*) causes severe defects in homologous recombination and recombinational repair in *Saccharomyces cerevisiae* (23), suggesting that these genes cooperatively function in DNA recombination. The human and *S. cerevisiae* Rad51 proteins, HsRad51 and ScRad51, respectively, catalyze homologous pairing reactions *in vitro*(24-26), thus indicating functional similarities between Rad51 and RecA.

Rad52 has been found as another member of the *RAD52* epistasis group in *S. cerevisiae*. Although the *RAD52* gene has been identified in yeast, mouse, chicken, and human, only the N-terminal half of the amino acid sequence of Rad52 is conserved in the proteins. It has been reported that Rad52 directly binds Rad51 and activates the Rad51-dependent homologous pairing reaction *in vitro*(21, 27). To investigate the molecular mechanisms of homologous pairing in the human, we analyzed the HsRad51 regions, which interact with the human Rad52 (HsRad52), by an NMR chemical shift perturbation experiment and GST-pull down assays using the HsRad51 mutants and an N-terminal peptide of HsRad51 (28, 29). Interestingly, the N-terminal region (1-114 amino acid residues, Fig. 2A) of HsRad51 did not exhibit strong binding to HsRad52, although the corresponding N-terminal region (1-152 amino acid residues) of ScRad51 has been suggested to bind ScRad52. On the other hand, we could isolate mutants that exhibited a significant decrease in HsRad52 binding from a pool of randomly generated C-terminal region mutants of HsRad51 (28). Among them, the Phe259 residue, which is strongly required for the HsRad52-binding, was a possible candidate for a direct HsRad52-binding site, because it is located on the molecular surface and is exposed to the solvent when the amino acid sequence of HsRad51 is superimposed on the crystal structure of RecA (Fig. 2B).

A　　　　　　　　**B**

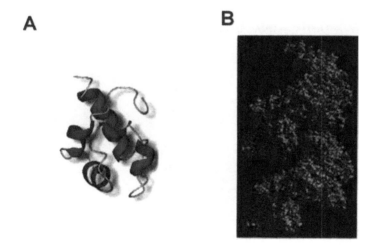

Figure 2 A, Solution structure of the N-terminal domain of HsRad51(1-114). B, Location of the Phe259 residue of HsRad51 superimposed on the crystal structure of RecA. The amino acid residue of RecA corresponding to Phe259 of HsRad51 is shown in red.

It has been reported that homologous pairing is enhanced when HsRad52 is incubated with ssDNA before the addition of HsRad51 and dsDNA. However, we found that HsRad52 itself promotes homologous pairing between ssDNA fragments and superhelical dsDNA (28). Interestingly, the homologous pairing activity of HsRad52 is apparently stronger than that of HsRad51 when superhelical dsDNA is used as a substrate. Further analysis is required to elucidate the biological significance of this finding.

2.2. DNA recombination activity of the human Xrcc3·Rad51C complex

The chromosomal DNA in the brain has a particularly high risk of DNA damage, including DSBs, because of the byproducts of extensive oxidative respiration. However, in mammals, the expression of *RAD51* mRNA, which is believed to be an essential gene product in the recombination repair of DSBs, is extremely low in the brain (22). Therefore, the brain might have a brain-specific activator of Rad51 or an unidentified functional homologue of Rad51. On the other hand, the human Xrcc3 protein, which appears to be a homologue of ScRad57, is expressed in the brain, as well as in the testis and the spleen, unlike the case of HsRad51 (30). The Xrcc3 protein is the product of one of the human X-ray repair cross complementing (*XRCC*) genes, which complement the hamster mutant cell phenotypes of extreme sensitivity to DNA damage and severe chromosome instability (30, 31). Recently, it has been reported that Xrcc3 is important to promote the repair of DSBs by homologous recombination and is associated with the development of melanoma skin cancer (32, 33). On the other hand, it has been found, by a two-hybrid analysis, that Xrcc3 interacts with the human Rad51C protein (Rad51C), which exhibits the highest sequence identity (27%) with HsRad51 (34).

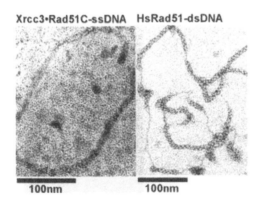

Figure 3 Electron microscopic visualizations of Xrcc3·Rad51C (left) and HsRad51 (right) complexed with single-stranded DNA and double-stranded DNA, respectively.

By a two-hybrid analysis using a human brain cDNA library, we revealed that the major Xrcc3-interacting protein is a Rad51 paralog, Rad51C, in the human brain (35). The purified Xrcc3·Rad51C complex, which shows apparent 1:1 stoichiometry, was found to catalyze the homologous pairing without HsRad51 (35). Although the activity is reduced, the Rad51C protein alone also catalyzed homologous pairing, suggesting that Rad51C is a catalytic subunit for homologous pairing. The DNA binding activity of Xrcc3•Rad51C was drastically decreased in the absence of Xrcc3, indicating that Xrcc3 is important for the DNA binding of Xrcc3•Rad51C. Electron microscopic observations revealed that Xrcc3•Rad51C formed filamentous structure with circular single-stranded DNA similar to HsRad51 (Fig. 3). Recently, Xrcc3 was suggested to be involved in the suppression of cancer. These results not only provide new insight into DNA repair in the human brain, but also are important to understand the mechanism of cancer cell proliferation.

2.3. Structure of a DNA-binding region of the human centromere protein B

The centromere plays a major role in chromosome segregation in mitosis and meiosis, by serving as the site for kinetochore assembly and chromatid attachment (36, 37). To understand the functions of the centromere at the molecular level, the DNAs and the proteins in centromeres have been characterized for a variety of eukaryotes (38). In human cells, several proteins localized in the centromere have been identified with centromere-specific autoantibodies. Among the centromere-specific proteins, CENP-A (17 kD), CENP-B (80 kD), and CENP-C (140 kD) are known to interact with DNA (39-42). Thus far, a sequence-specific DNA binding activity has been found only for CENP-B.

Figure 4 Structure of CENP-B domain 1. Residues forming the hydrophobic core are shown in red.

CENP-B binds a 17 base pair DNA sequence (CENP-B box) in the N-terminal region from 1 to 129 amino acid residues. We determined the structure of the CENP-B polypeptide from 1 to 56 amino acid residues (43), and revealed that four α-helices exist in this region (Fig. 4). This domain (domain 1) contains a canonical helix-turn-helix motif. The chemical shift perturbation experiment showed that domain 1 interacted with one half of CENP-B box sequence. Currently, we are determining the crystal structure of CENP-B$_{1-129}$, which contains amino acid residues 1 to 129, complexed with the CENP-B box DNA (Fig. 5) (44). The solution structure of CENP-B$_{1-56}$ was essentially the same as that of the CENP-B$_{1-129}$•DNA complex.

3. ANALYSES OF RAS AND ITS RELATED PROTEINS IN SIGNAL TRANSDUCTION

The Ras proteins, 21 kDa low molecular mass guanine nucleotide-binding proteins, are important for cell proliferation, terminal differentiation, and other cellular events (45). The Ras protein is known to interact with many proteins, including its activator and target proteins, also indicating its indispensability in cells. In this section, we focus on the structural biology of Ras and its related proteins, which are involved in the signal transduction pathway.

3.1. Structural analyses of the c-Ha-Ras protein and its guanine nucleotide exchange factor, the Sos protein

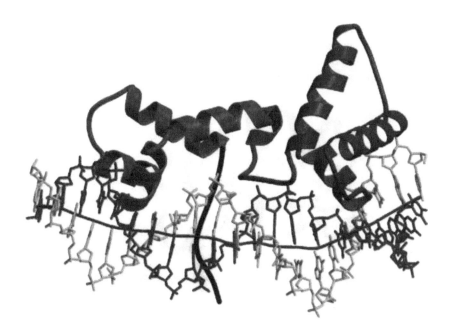

Figure 5 The crystal structure of the CENP-B$_{1-129}$ protein complexed with the 21-mer CENP-B box DNA. DNA strands are shown in yellow and green. The N-terminal loop, domain 1, linker loop, and domain 2 of the CENP-B$_{1-129}$ protein are shown in blue, orange, pink, and light blue, respectively.

The Ras protein is tightly bound to GDP or GTP (45). The GDP-bound Ras protein is predominant in non-stimulated cells, whereas the GTP-bound Ras protein is present in increased amounts in cells stimulated by proliferation or differentiation signals (46-50). This indicates that the GTP-bound Ras protein is the active form. GMPPNP was used to mimic the GTP-bound form of Ras, and the Ras·GMPPNP complex was analyzed by multidimensional NMR spectroscopy (51). In the NMR analysis, the backbone amide resonances of amino acid residues 10-13, 21, 31-39, 57-64, and 71 of Ras(1-171) ·GMPPNP, but not those of Ras(1-171) ·GDP, were extremely broadened. These broadenings of the amide resonances were ascribed to the exchange at an intermediate rate on the NMR time scale. The residues exhibiting extreme broadening, except for residues 21 and 71, are localized in three functional loop regions, loops L1, L2 (switch I), and L4 (switch II). These functional loops are involved in GTP hydrolysis and interactions with other proteins. This is a characteristic feature of the GTP-bound form of Ras, in which the L1, L2, and L4 loop regions, but not the other regions, rather slowly interconvert between two or more stable conformers. This phenomenon, termed "regional polysterism", exhibited by these loop regions may be related to their multifunctionality.

The formation of GTP-bound Ras is promoted by guanine nucleotide exchange factors (GEFs), such as the mouse and human Son-of-sevenless proteins (mSos and hSos, respectively)(52, 53). The solution structure of the pleckstrin homology (PH) domain of mSos1 was determined by multidimensional NMR spectroscopy (Fig. 6A)(54). The structure of the mSos1 PH domain involves the fundamental PH fold, consisting of seven β-strands and one α-helix at the C-terminus, as determined for the PH domains of other proteins. The mSos1 PH domain was found to bind phosphatidylinositol-4,5-bisphosphate by a centrifugation assay. The addition of inositol-1,4,5-trisphosphate to the mSos1 PH domain induced backbone amide chemical shift changes, mainly in the β1/β2 loop and the β3/β4 loop, which are characteristically unstructured in the mSos1 PH domain.

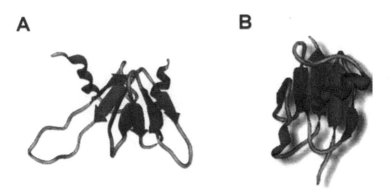

A **B**

Figure 6 Solution structures of the mSos1 PH domain (A) and RGL-RBD (B).

3.2.Structural basis for interactions of Ras or Rap1A with its target proteins

The Ras protein and its homolog, Rap1A, have an identical "effector region" (residues 32-40) preceded by Asp30-Glu31 and Glu30-Lys31, respectively (55). In the complex of the "Ras-like" E30D/K31E mutant Rap1A with the Ras-binding domain (RBD), residues 51-131 of Raf-1, Glu31 in Rap1A forms a tight salt bridge with Lys84 in Raf-1. However, we have found that the Raf-1 RBD binding of Ras is indeed reduced by the E31K mutation, but is not affected by the E31A mutation (56).   The "Rap1A-like" D30E/E31K mutant of Ras was shown to bind the Raf-1 RBD less strongly than wild-type Ras, but slightly more tightly than the E31K mutant.   The NMR analysis revealed that the Lys84 residue in the Raf-1 RBD interacts with wild-type Ras (57).   The D30E/E31K mutant of Ras caused nearly the same perturbations of Raf-1 chemical shifts, including Lys84.   Therefore, we hypothesized that Glu31 in Ras may not be the major salt bridge partner of Lys84 in Raf-1.   The Ras Asp33 residue, which is required for the Raf-1 binding, may be the major salt bridge partner of Lys84 in Raf-1.

The RGL protein, a homolog of the Ral GDP dissociation stimulator (RalGDS), has been identified as a downstream effector of Ras (58-62).   The solution structure of the Ras-binding domain of RGL (RGL-RBD) was determined by NMR spectroscopy (Fig. 6B)(63). The overall fold of RGL-RBD consists of a five-stranded β-sheet and two α-helices, which is the same topology as that of RalGDS-RBD.   Chemical shift perturbation experiments showed that the Ras-interacting residues of RGL-RBD are appreciably different from those of RalGDS-RBD.

REFERENCES

1. R. Giegé, M. Sissler and C. Florentz, Nucleic Acids Res., 26 (1998) 5017.
2. A. Fersht, in Enzyme structure and mechanism, second edition W. H. Freeman and Company, New York, pp.1-1475 (1985).
3. O. Nureki, D. G. Vassylyev, M. Tateno, A. Shimada, T. Nakama, S. Fukai, M. Konno, T. L. Hendrickson, P. Schimmel and S. Yokoyama, Science, 280 (1998) 578.
4. S. Fukai, O. Nureki, S. Sekine, A. Shimada, J. Tao, D. G. Vassylyev and S. Yokoyama, Cell, 103 (2000) 793.
5. J. Lapointe, L. Duplain and M. Proulx, J. Bacteriol., 165 (1986) 88.
6. A. Schön, G. Kannangara, S. Gough and D. Söll, Nature, 331 (1988) 187.
7. K. C. Rogers and D. Söll, J. Mol. Evol., 40 (1995) 476.
8. Y. Gagnon, L. Kacoste, N. Champagne and J. Lapointe, J. Biol. Chem., 271 (1996) 14856.
9. A. W. Curnow, D. L. Tumbula, J. T. Pelaschier, B. Min and D. Söll, Proc. Natl. Acad. Sci. USA, 95 (1998) 12838.
10. H. D. Becker and D. Kern, Proc. Natl. Acad. Sci. USA, 95 (1998) 12832.
11. J. Handy and R. F. Doolittle, J. Mol. Evol., 49 (1999) 709.
12. S. Sekine, O. Nureki, A. Shimada, D. G. Vassylyev and S. Yokoyama, Nature Struct. Biol., 8 (2001) 203.
13. J. B. Harford and D. R. Morris, mRNA metabolism and post-transcriptional gene regulation Wiley-Liss, New York, 1997.
14. T. W. Cline and B. J. Meyer, Annu. Rev. Genet., 30 (1996) 637.
15. B. A. Sosnowski, J. M. Belote and M. McKeown, Cell, 58 (1989) 449.
16. K. Inoue, K. Hoshijima, H. Sakamoto and Y. Shimura, Nature, 344 (1990) 461.
17. J. Valcárcel, R. Singh, P. D. Zamore and M. R. Green, Nature, 362 (1993) 171.
18. N. Handa, O. Nureki, K. Kurimoto, I. Kim, H. Sakamoto, Y. Shimura, Y. Muto and S. Yokoyama, Nature, 398 (1999) 579.
19. T. Shibata, C. DasGupta, R.P. Cunningham and C.M. Radding, Proc. Natl. Acad. Sci.

USA, 76 (1979) 1638.

20. K. McEntee, G. M. Weinstock and I. R. Lehman, Proc. Natl. Acad. Sci. USA, 76 (1979) 2615.

21. A. Shinohara, H. Ogawa and T. Ogawa, Cell, 69 (1992) 457.

22. A. Shinohara, H. Ogawa, Y. Matsuda, N. Ushio, K. Ikeo and T. Ogawa, Nature Genet., 4 (1993) 239.

23. T. D. Petes, R. E. Malone and L. S. Symington, in The molecular and cellular biology of the yeast Saccharomyces (J. R. Broach, J. R. Pringle and E. W. Jones eds), Cold Spring Harbor Laboratory Press, Cold Spring Harbor, New York, pp.407 (1991).

24. P. Sung, Science, 265 (1994) 1241.

25. P. Baumann, F. E. Benson and S. C. West, Cell, 87 (1996) 757.

26. R. C. Gupta, L. R. Bazemore, E. I. Golub and C. M. Radding, Proc. Natl. Acad. Sci. USA, 94 (1997) 463.

27. Z. Shen, K. G. Cloud, D. J. Chen and M. S. Park, J. Biol. Chem., 271 (1996) 148.

28. H. Kurumizaka, H. Aihara, W. Kagawa, T. Shibata and S. Yokoyama, J. Mol. Biol., 291 (1999) 537.

29. H. Aihara, Y. Ito, H. Kurumizaka, S. Yokoyama and T. Shibata, J. Mol. Biol., 290 (1999) 495.

30. N. Liu, J. E. Lamerdin, R. S. Tebbs, D. Schild, J. Tucker, M. R. Shen, K. W. Brookman, M. J. Siciliano, C. A. Walter, W. Fan, L. S. Narayana, Z. Zhou, A. W. Adamson, K. J. Sorensen, D. J. Chen, N. J. Jones and L. H. Thompson, Mol. Cell, 1 (1998) 783.

31. R. S. Tebbs, Y. Zhao, J. Tucker, J. B. Scheerer, M. J. Siciliano, M. Hwang, N. Liu, R. J. Legerski and L. H. Thompson, Proc. Natl. Acad. Sci. USA, 92 (1995) 6354.

32. A. J. Pierce, R. D. Johnson, L. H. Thompson and M. Jasin, Genes Dev., 13 (1999) 2633.

33. S. L. Winsey, N. A. Haldar, H. P. Marsh, M. Bunce, S. E. Marshall, A. L. Harris, F. Wojnarowska and K. I. Welsh, Cancer Res., 60 (2000) 5612.

34. M. K. Dosanjh, D. W. Collins, F. Wufang, G. G. Lennon, J. S. Albala, Z. Shen and D. Schild, Nucleic Acids Res., 26 (1998) 1179.

35. H. Kurumizaka, S. Ikawa, M. Nakada, K. Eda, W. Kagawa, M. Takata, S. Takeda, S. Yokoyama and T. Shibata, Proc. Natl. Acad. Sci. USA, 98 (2001) 5538.

36. L. Clarke, Trends Genet., 6 (1990) 150.

37. H. F. Willard, Trends Genet., 6 (1990) 410.

38. A. F. Plûta, A. M. Mackay, A. M. Ainsztein, I. G. Goldberg and W. C. Earnshaw, Science, 280 (1995) 1591.

39. H. Masumoto, H. Masukata, Y. Muro, N. Nozaki and T. Okazaki, J. Cell. Biol., 109 (1989) 1963.

40. D. K. Palmer, K. O'Day, H. L. Trong, H. Charbonneau and R. L. Margolis, Proc. Natl. Acad. Sci. USA, 88 (1991) 3734.

41. K. Sugimoto, H. Yata, Y. Muro and M. Himeno, J. Biochem. (Tokyo), 116 (1994) 877.

42. K. F. Sullivan, M. Hechenberger and K. Masri, J. Cell. Biol., 127 (1994) 581.

43. J. Iwahara, T. Kigawa, K. Kitagawa, H. Masumoto, T. Okazaki and S. Yokoyama, EMBO J,
17 (1998) 827.

44. Y. Tanaka, O. Nureki, H. Kurumizaka, S. Fukai, S. Kawaguchi, M. Ikuta, J. Iwahara, T. Okazaki and S. Yokoyama, EMBO J., in press.

45. M. Barbacid, Annu. Rev. Biochem., 56 (1987) 779.

46. T. Satoh, S. Nakamura and Y. Kaziro, Mol. Cell. Biol., 7 (1987) 4553.

47. T. Satoh, M. Endo, M. Nakafuku, T. Akiyama, T. Yamamoto and Y. Kaziro, Proc. Natl. Acad. Sci. USA, 87 (1990) 7926.

48. T. Satoh, M. Endo, M. Nakafuku, S. Nakamura and Y. Kaziro, Proc. Natl. Acad. Sci. USA, 87 (1990) 5993.

49. T. Satoh, M. Nakafuku, A. Miyazima and Y. Kaziro, Proc. Natl. Acad. Sci. USA, 88 (1991) 3314.

50. M. Trahey and F. McCormick, Science, 248 (1987) 542.

51. Y. Ito, K. Yamasaki, J. Iwahara, T. Terada, A. Kamiya, M. Shirouzu, Y. Muto, G. Kawai, S. Yokoyama, E. D. Laue, M. Wälchli, T. Shibata, S. Nishimura and T. Miyazawa, Biochemistry, 36 (1997) 9109.

52. D. Bowtell, P. Fu, M. Simon and P. Senior, Proc. Natl. Acad. Sci. USA, 89 (1992) 7100.

53. P. Chardin, J. H. Camonis, N. W. Gale, L. van Aelst, J. Schlessinger, M. H. Wigler and D. Bar-Sagi, Science, 260 (1993) 1388.

54. S. Koshiba, T. Kigawa, J-H. Kim, M. Shirouzu, D. Bowtell and S. Yokoyama, J. Mol. Biol., 269 (1997) 579.

55. V. Pizon, P Chardin, I. Lerosey, B. Olofsson and A. Tavitian, Oncogene, 3 (1988) 201.

56. M, Shirouzu, K. Morinaka, S. Koyama, C-D. Hu, N. Hori-Tamura, T. Okada, K-i. Kariya, T. Kataoka, A. Kikuchi and S. Yokoyama, J. Biol. Chem., 273 (1998) 7737.

57. T. Terada, Y. Ito, M. Shirouzu, M. Tateno, K. Hashimoto, T. Kigawa, T. Ebisuzaki, K. Takio, T. Shibata, S. Yokoyama, B. O. Smith, E. D. Laue and J. A. Cooper, J. Mol. Biol., 286 (1999) 219.

58. F. Hofer, S. Fields, C. Schneider and G. S. Martin, Proc. Natl. Acad. Sci. USA, 91 (1994) 11089.

59. A. Kikuchi, S. D. Demo, Z.-H. Ye, Y.-W. Chen and L. T. Williams, Mol. Cell. Biol., 14 (1994) 7483.

60. M. Spaargaren and J. R. Bischoff, Proc. Natl. Acad. Sci. USA, 91 (1994) 12609.

61. S. N. Peterson, L. Trabalzini, T. R. Brtva, T. Fischer, D. L. Altschuler, P. Martelli, E. G. Lapetina, C. J. Der and G. C. White II, J. Biol. Chem., 271 (1996) 29903.

62. R. M. F. Wolthuis, B. Bauer, L. J. van't Veer, A. M. M. de Vries-Smith, R. H. Cool, M Spaargaren, A. Wittinghofer, B. M. T. Burgering and J. L. Bos, Oncogene, 13 (1996) 353.

63. T. Kigawa, M. Endo, Y. Ito, M. Shirouzu, A. Kikuchi and S. Yokoyama, FEBS letters, 441 (1998) 413.

Molecular Anatomy of Cellular Systems
I. Endo et al., (editors)

The importance of the hydrophobic pocket in actin subdomain 4 for Ca^{2+}-activation of actin-activated myosin ATPase in the presence of Tropomyosin-Troponin

Takeyuki Wakabayashi[a], Takuo Yasunaga[a], Kimiko Saeki[a], Yoshiyuki Matsuura[a]

[a] Department of Physics, School of Science, University of Tokyo
Hongo 7-3-1, Bunkyo-ku, Tokyo 113-0033, Japan

We have generated actin mutants to probe the mechanism by which Ca^{2+} activates muscle contraction through tropomyosin and troponin. To prepare functional mutant actins, we used the system of *Dictyostelium discoideum*. First, we generated a chimeric actin, in which the sequence 228-232 (QTAAS) of *Dictyostelium* actin was replaced with that of *Tetrahymena* actin (KAYKE). The initial intension was to determine tropomyosin-binding sites, because *Tetrahymena* actin does not bind tropomyosin and two actins are different in five consecutive amino acids. As expected the chimeric actin showed poorer tropomyosin binding, and it was concluded that the sequence 228-232, which is in the subdomain 4 of actin, is an important tropomyosin-binding site. Interestingly, the chimeric actin showed unexpected "higher Ca^{2+}-activation" of myosin ATPase in the presence of tropomyosin-troponin with Ca^{2+} than the wild-type actin, even though the chimeric actin *per se* showed the normal activation of myosin ATPase. Later, it was found that this "higher Ca^{2+}-activation" is solely due to the replacement of alanine 230 by tyrosine (A230Y mutant). Because the A230Y-mutant actin showed normal tropomyosin binding, the "higher Ca^{2+}-activation" is not the consequence of poorer binding of tropomyosin. X-ray crystallographic study showed that the overall main chain structure of chimeric actin was almost the same as that of wild-type actin, whereas the side-chain of leucine 236, which was not mutated, was displaced in mutant actins so that the hydrophobic pocket in the subdomain 4 became more accessible. We proposed therefore that actin would activate myosin ATPase more strongly when tropomyosin binds to this pocket. To confirm this proposal, we truncated the side chain of leucine 236 (A230Y/L236A-mutant) to further increase the accessibility of the hydrophobic pocket. It was found that such a mutant showed much higher Ca^{2+}-activation in the presence of tropomyosin. Unlike the previous A230Y-mutants, the A230Y/L236A-mutant actin did not require the co-existence of troponin and Ca^{2+} for the higher activation. Therefore, we concluded that tropomyosin would bind to the hydrophobic pocket more firmly when A230Y/L236A-mutation was introduced and such actin-tropomyosin complexes activated myosin ATPase much more strongly than wild-type actin.

1. INTRODUCTION

The binding of calcium ions to troponin C triggers vertebrate striated (skeletal or cardiac) muscle contraction through a series of interactions involving the regulatory proteins including tropomyosin and troponin that regulate the interaction between actin and myosin (1) that ultimately generates force by sliding (2, 3). However, the nature of these interactions and the precise mechanisms by which they control contraction has been not completely understood. The modeling based on X-ray scattering has given an indication of how tropomyosin is relocated during activation (4-6). Three-dimensional electron microscopy has shown that tropomyosin binds to the inner domain of actin, which corresponds to subdomain 3 and 4, in the presence of Ca^{2+} ("on" state) (7-12). Although electron microscopy and the modeling based on X-ray scattering have given an indication of tropomyosin shift, clear three-dimensional images have not yet been obtained with sufficiently high resolution to enable the precise binding site for tropomyosin to be established. In the analysis of electron micrographs or the modeling based on X-ray scattering data, it has been assumed that tropomyosin would follow the helical symmetry of actin and that troponin that does not follow the helical symmetry would not affect the results. It was also pointed out that these assumptions might not be correct especially in the absence of Ca^{2+} (13). Narita et al. therefore reconstructed three-dimensional images of reconstituted thin filaments using a single particle method without assuming the helical symmetry (14). They found that tropomyosin at low Ca^{2+} did not follow the helical symmetry and that the Ca^{2+}-induced shift of troponin affected the positions of tropomyosin that were determined by helical reconstruction.

Although the primary structure of actin is highly conserved across species, *Tetrahymena* actin sequence is comparatively divergent and cannot bind either tropomyosin or phalloidin (15-17). Residues 228-232 in the subdomain 4 of actin are the most different region between *Tetrahymena* actin and conventional actins including skeletal actin and cytosolic actin from *Dictyostelium*, except for N-terminus region where identity is generally low among species. The actin chimera, in which the *Dictyostelium* sequence was replaced by the corresponding sequence of *Tetrahymena* (QTAAS in the sequence of 228-232 to KAYKE replacement), binds tropomyosin poorly as expected, but surprisingly, it shows a higher ratio of the activation of the myosin S1 ATPase in the presence of tropomyosin-troponin and Ca^{2+} to that in their absence in comparison with that of the wild-type actin (18, 19). We refer this effect to "higher Ca^{2+}-activation". To investigate its mechanism various mutant actins were generated and some of them were studied using X-ray crystallography (20).

2. Ca^{2+}-ACTIVATION OF S1-ATPASE IN THE PRRSENCE OF TROPOMYOSIN-TROPNIN

When the activation of myosin ATPase is measured at low myosin concentrations, activation by actin-tropomyosin-troponin at high Ca^{2+} is much lower than that by pure F-actin. Either allosteric models or three-state models (21) can explain this apparent suppression by tropomyosin-troponin. In the former model, there is equilibrium between a T-state and an R-state at high Ca^{2+}, with a T-state being more favorable at low myosin concentrations.

2.1. The role of the hydrophobic pocket in the subdomain 4 of actin

Figure 1 shows the ribbon model representation of wild-type actin from *Dictyostelium* (20). It can be seen that the crystal structure of *Dictyostelium* actin is almost the same as that of rabbit skeletal actin (17). Altered five residues in chimeric actin are shown in ball-and-stick in the ribbon model. The ability of the mutant actins to activate the myosin S1-ATPase under four conditions, i. e., pure actin, in the presence of tropomyosin, in the presence of tropomyosin-troponin either with or without Ca^{2+} was examined (18, 19). Without regulatory proteins, all of the mutant actins activated the S1-ATPase to the levels comparable to those observed with wild-type actin. In the

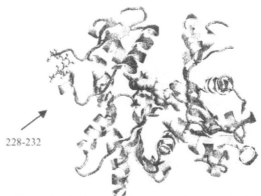

Figure 1. Ribbon model representation of the wild-type actin from *Dictyostelium* (20) showing the positions of site-directed mutagenesis by ball-and-stick model and indicated by an arrow. Bound ATP is shown by space-fill model.

presence of tropomyosin, the activation by actin from half chimera-2 with the sequence of QTA**K**E (mutated amino acids are represented in bold face) or Q228K was significantly lower than that by other mutants or the wild-type actin. When both of tropomyosin and troponin were added in the absence of Ca^{2+}, the activation by every actin mutant was suppressed in the same way as wild-type actin, indicating that the inhibition of the S1-ATPase by tropomyosin-troponin from rabbit skeletal muscle was functioning normally. However, in the presence of both tropomyosin-troponin and Ca^{2+}, there were distinctive differences between the various mutants: The activation of the myosin S1-ATPase observed for half chimera-2 was similar to that observed for wild-type actin but the half chimera-1 with the sequence of **K**A**Y**AS showed a substantially (~3 times) higher Ca^{2+}-activation of the S1-ATPase than the wild-type actin. Thus the half chimera-1 has retained the property of the "higher Ca^{2+}-activation" observed with the full chimera with the sequence of **K**A**YK**E (mutated amino acids are represented in bold face).

To examine the contributions made by individual residues in the half chimera-1 to the "higher Ca2+-activation", point mutants Q228K and A230Y were investigated. Because T229C actin showed similar properties to the wild-type actin, it was anticipated that the introduction of K or Y would induce the "higher Ca^{2+}-activation". The A230Y mutant was found to show the "higher Ca^{2+}-activation" as observed with the full chimera, whereas Q228K actin behaves as the wild-type actin. These results indicate that introduction of tyrosine at position 230 is sufficient for the "higher Ca^{2+}-activation".

To examine the extent of "higher Ca^{2+}-activation", the ratio of the activation of myosin S1 ATPase by actin in the presence of tropomyosin-troponin and Ca^{2+} to that in their absence was calculated. This ratio (normalized ATPase activation) represents the effect of mutagenesis on the regulatory mechanism through the interaction of actin with tropomyosin-troponin, because the effect of mutagenesis on the activation by pure actin is eliminated by normalization. The mutant actins can be classified into two groups: The actins from half chimera-2 and Q228K show similar normalized activation to that by the wild-type actin (~0.3), whereas A230Y, half chimera-1 and chimera show two to three times higher normalized activation (from ~1.0 to ~1.3) than the wild-type. Because the common feature of the latter group is that the residue 230 was mutated to tyrosine, we refer to them as "Tyr"-mutants. This indicates that the region around the residue 230 plays an important role in Ca^{2+}-activation by actin-tropomyosin-troponin.

2.2. Proposed Mechanism of "Higher Ca^{2+}-Activation" of myosin ATPase by Tyr-mutant actins

Without tropomyosin-troponin all of the four *Dictyostelium* mutant actins activated ATPase as the wild-type actin and this activation was suppressed in a normal manner by rabbit skeletal tropomyosin-troponin in the absence of Ca^{2+}. This indicates that the basic process of actin activation and the inhibitory process in mutant actin systems are undistinguishable from that of the wild-type one. However, Tyr-mutants containing A230Y mutation showed the "higher Ca^{2+}-activation". This indicates that only the Ca^{2+}-activation process is affected. The atomic structure of the main chain of the half chimera-1 with the **KAYAS** sequence is almost the same as that of the wild-type actin as determined by X-ray (20). However, it was noticed that the side chain of leucine 236 was displaced in the mutant actins so that the hydrophobic pocket of the subdomain 4 of actin became more accessible to solvent as shown in Figure 2. Therefore, it is plausible that the effect of mutagenesis on the activation of myosin ATPase is not due to the structural changes of actin *per se* but through the changes in the interaction of actin with tropomyosin and/or troponin. Figure 3 shows that the residue 230 shown in black that locates in the hydrophobic pocket can be scarcely seen

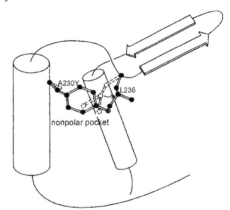

Figure 2. Schematic model of the changes in the subdomain 4 of actin induced by Ala230 to Tyr mutation. The side chain of leucine 236 swung away and the hydrophobic pocket in the subdomain 4 became more exposed and the Tyr230 of the mutant actin more accessible than Ala230 of the wild-type actin. We propose that actin-activated myosin ATPase increase almost 10 times, when tropomyosin binds to the nonpolar pocket.

in the wild-type actin (Fig. 3A) but it can be seen in the half-chimera actin (Fig. 3B). In the wild-type actin, the access to the residue 230 is almost blocked by the side chains of residues 229, 235 and 236 shown in dark gray, which are located in the left, top and right of the residue 230, respectively. In the half-chimera actin, the residue 229 is mutated from threonine to alanine and the side chain is less bulky, the side chains of residues 235 and 236 were displaced so that the pocket became more accessible to the solvent (Fig. 3B). The residues 229, 235, and 236 therefore seem to play an important role of a gatekeeper to restrict the access to the pocket. Knowing that the position of the side chain of leucine 236 is important, we generated the double mutant: Leucine 236 was mutated to alanine in addition to the mutation of the half chimera-1 to ensure increased accessibility of the hydrophobic pocket to the solvent. It was found that such mutant showed the even higher Ca^{2+}-activation in the presence of tropomyosin-troponin at high Ca^{2+}. It was also found that the similar activation could be achieved by tropomyosin only (22). Therefore, the binding of tropomyosin to the pocket seems to play a major role in the "higher activation". We propose that actin would activate myosin ATPase most strongly when tropomyosin binds to the hydrophobic pocket of the subdomain 4 of actin. When Ca^{2+} concentration is low tropomyosin locates far from hydrophobic pocket, while it shifts near to the hydrophobic pocket at high Ca^{2+}. However, in the case of the wild-type actin, the accessibility of hydrophobic pocket for tropomyosin is limited and actin cannot activate myosin fully unless myosin concentration is very high. When the hydrophobic pocket became more readily accessible to the solvent, tropomyosin can bind to the hydrophobic pocket and actin can activate myosin ATPase fully even when myosin concentration is not high.

It is possible to interpret our data according to an allosteric model or three-state model (21). We propose that actin from Tyr-mutant favors an R-state ("open state"), and push the T-R equilibrium from a T-state ("closed state") towards an R-state ("open-state"). Because the ATPase activation by the Tyr-mutant actins in the presence of Ca^{2+}-tropomyosin-troponin is 2-3 times higher than that without tropomyosin-troponin, the role of Ca^{2+} is not just suppressing the inhibition by troponin. This kind of "higher Ca^{2+}-activation" of ATPase activation cannot be explained by the simple steric blocking. Thus, the combination of the steric blocking (5-12) and an allosteric/cooperative model such as a three-state model is needed to explain the results of our mutagenesis experiments and many structural studies on thin filaments.

The increase of Ca^{2+} concentration causes a conformational changes of troponin (C+I) (9) or troponin(T+I+C) (12), which is transmitted to actin through tropomyosin. The local changes in the side chain conformation caused by introduction of A230Y mutagenesis may alter the interactions between actin and tropomyosin and/or troponin. We proposed that Tyr-mutant actin favors an R-state ("open state") through the stronger binding of tropomyosin to the hydrophobic pocket in the subdomain 4 of actin at high Ca^{2+}. The conformation of the wild-type and Tyr-mutant actins should be almost the same in the absence of

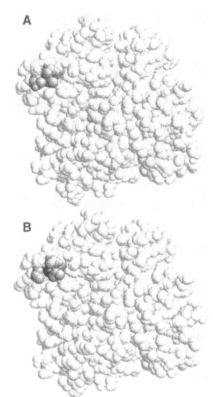

Figure 3. Solid models of (A) the wild-type with the sequence of QTAAS and (B) the half-chimera actin with the sequence of KAYAS. The residue 230 is black, and residues 229(left), 235 (top) and 236 (right) are in dark gray. The residue 230 (black) is visible in the half-chimera actin (B) but can scarcely be seen in the wild-type actin (A).

Ca^{2+}. However, an equilibrium constant K_T or $1/L$ (between R-state ("open state") and T-state ("closed state")) of "Tyr"-mutant actins may be larger than that of wild-type or "non-Tyr"-mutant ones so that an R-state ("open-state") is more favored at high Ca^{2+}.

3. TROPOMYOSIN-BINDING TO ACTIN MUTANTS

It has been proposed that charged amino acid residues are important for actin-tropomyosin interaction in both the "on" and "off" states. The distribution of charged amino acid residues along tropomyosin molecules shows 14 repeats. Elimination of two charged residues from a *Dictyostelium* actin (K238A/E241A) results in poorer tropomyosin binding (18). On the surface of tropomyosin molecule there are a large number of residues such as serine, asparagine, threonine, or glutamine which can form hydrogen bonds and the hydroxyl groups of the QTAAS sequence on actin may form a number of hydrogen bonds with tropomyosin that supplement the electrostatic interactions between these molecules. The hydrogen bonds might be disturbed by the introduction of large charged side chain of Lys and Glu to the last two positions of the pentapeptide. It is important to discriminate the interactions between actin and tropomyosin in a T-state and that in an R-state. When the binding of tropomyosin to actin is assayed by co-pelleting, the interactions in a T-state are measured. The interaction became weaker with K238A/E241A mutation as described. However, the interaction at an R-state would depend more on the accessibility of the hydrophobic pocket in the subdomain 4 to the solvent so that tropomyosin can bind to it. When tropomyosin binds to the hydrophobic pocket of actin then actin-tropomyosin-troponin would activate myosin ATPase more fully.

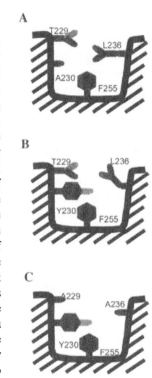

Figure 4. Schematic representation of the hydrophobic pocket of (A) wild-type actin (B) half-chimera 1, and (C) half-chimera 1 and L236A.

4. SUMMARY

In summary, the introduction of tyrosine replacing alanine 230 induced the "higher Ca^{2+}-activation". The "higher Ca^{2+}-activation" is not the consequence of poorer tropomyosin binding, because Ala230-to-Tyr mutant actin bound tropomyosin normally. The higher activation by Tyr-mutant actin may be caused by shifting allosteric T-R equilibrium towards an R-state ("open state") through the stronger binding of tropomyosin to the hydrophobic pocket in the subdomain 4 of actin. These mutant actins will help to elucidate the regulatory mechanism of muscle contraction.

REFERENCES

1. S. Ebashi and M. Endo, Progr. Biophys. Mol. Biol., 18 (1968) 123.
2. A. F. Huxley and R. Niedergerke, Nature (London), 173 (1954) 971.
3. H. E. Huxley and J. Hanson, Nature (London), 173 (1954) 973.
4. R. A. Milligan, and P. F. Flicker, J. Cell Biol., 105 (1987) 29.
5. T. Wakabayashi, H. E. Huxley, L. A. Amos and A. Klug, J. Mol. Biol., 93 (1975) 477.
6. T. Ishikawa, and T. Wakabayashi, Biochem. Biophys. Res. Commun., 203 (1994) 951.
7. W. Lehman, R. Craig and P. Vibert, Nature(London), 368 (1994) 65.
8. P. Vibert, R. Craig, and W. Lehman, J. Mol. Biol., 266 (1997) 8.
9. T. Ishikawa, and T. Wakabayashi, J. Biochem., 126 (1999) 200.
10. J. C. Haselgrove, Cold Spring Harbor Symp. Quant. Biol., 37 (1972) 341.
11. H. E. Huxley, Cold Spring Harbor Symp. Quant. Biol., 37 (1972) 361.
12. D. A. D. Parry, and J. M. Squire, J. Mol. Biol., 75 (1973) 33.
13. H. A. AL-Khayat, N. Yagi, and J.M. Squire, J. Mol. Biol., 252 (1995) 611.
14. A. Narita, T. Yasunaga, T. Ishikawa, K. Mayanagi and T. Wakabayashi, J. Mol. Biol., 308(2001) 241
15. M. Hirono, H. Endoh, N. Okada, O. Numata, and Y. Watanabe, J. Mol. Biol., 194 (1987) 181.
16. P. Sheterline, J. Clayton, and J. Sparrow, in Protein Profile: Actins (3rd Ed.) Academic Press Inc., San Diego (1996).
17. K. M. Flaherty, D. M. McKay, W. Kabsch, and K. C. Holmes, Proc. Natl. Acad. Sci. USA, 88 (1991) 5041.
18. K. Saeki, K. Sutoh and T. Wakabayashi, Biochemistry, 35 (1996) 14465.
19. K. Saeki and T. Wakabayashi, Biochemistry, 39 (2000) 1324
20 Y. Matsuura, M. Stewart, N. Kamiya, M. Kawamoto, K. Saeki, T. Yasunaga and T. Wakabayashi, J. Mol. Biol., 296 (2000) 579
21. D. F. McKillop and M. A. Geeves, Biophys. J., 65 (1993) 693.

Molecular Anatomy of Cellular Systems
I. Endo et al., (editors)
© 2002 Elsevier Science B.V. All rights reserved.

Physiological functions and molecular structures of new types of hemoproteins

Yoshitsugu Shiro[a], Yasuhiro Isogai[a], Hiro Nakamura[a] and Tetsutaro Iizuka[b]

[a] Biophysical Chemistry Laboratory, RIKEN Harima Institute/SPring-8, Kouto 1-1-1, Mikazuki-cho, Sayo, Hyougo 671-5143
[b] Department of Materials Chemistry, Faculty of Engineering, Hosei University, Koganei, Tokyo 184-8584, Japan

Phagocitic NADPH oxidase, NO reductase and O_2 sensor protein FixL are heme-containing proteins (hemoproteins), which are newly found in the past decade. They have been studied by spectroscopic, kinetic, crystallographic, biochemical and molecular biological techniques, to understand their pysiological functions on the basis of their molecular structures.

1. INTRODUCTION

A variety of transition metals are contained in many proteins as their active sites, and play crucial roles for physiological actions of these metal-binding proteins (metalloproteins and metal-enzymes). Two distinctly different functions can be mainly identified for the transition metals in metalloproteins. In one case, the metals can provide a binding site of some small molecules such as O_2, CO, NO, ethylene, and so on. In another case, the metals act as a redox or acid-base center of the catalyst (enzyme) in some biological reactions.

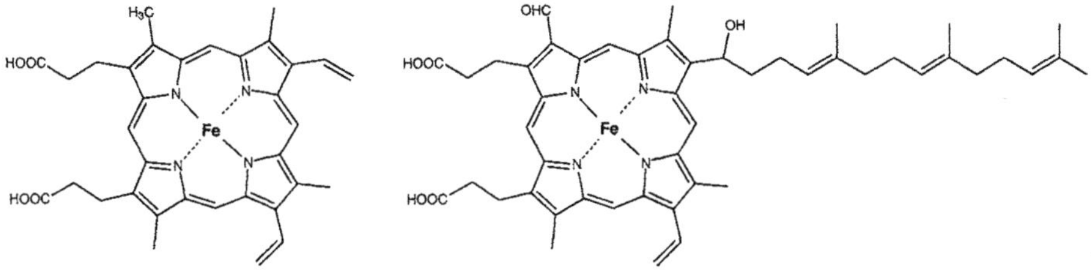

Figure 1. Structures of heme A and heme B

Hemoproteins are proteins/enzymes containing heme (Fe-porphyrin complex; see Figure 1) as a prosthetic group. A variety of hemoproteins such as myoglobin, hemoglobin, cytochromes, peroxidases, catalases, oxidases, oxygenases, are represented in biochemical textbooks. These can be also classified into the two main groups stated above. The typical examples of the former case are myoglobin and hemoglobin, both of which are O_2 carriers. A conformational change induced by the O_2 binding to the iron of one subunit of hemoglobin is highly related to the allosteric effects in the O_2 carrying function in blood. Cytochrome P450 is an example of the latter case, in which O_2 is activated at the iron site of the enzyme using the electrons from NAD(P)H, and physiologically responsible to the drug metabolism and steroid hormone synthesis.

In past decade, in addition to these classical hemoproteins, some hemoproteins are newly found in several organisms, which can be also classified into the two classes. Gas sensing hemoproteins such as CO sensor (CooA), O_2 sensor (FixL) and NO sensor (soluble guanylate cyclase) belong to the first class, in which protein conformational changes upon the ligand (CO, O_2, NO) binding to the heme iron are crucial for activation of catalytic sites which are located at different site of the proteins. NADPH oxidase, NO synthase (NOS) and NO reductase are redox hemproteins (second class), which produce O_2^- from O_2, NO from arginine and O_2, and N_2O from NO, rspectively.

In this text, we describe recent results of our studies on molecular mechanisms and structures of these newly found hemoproteins, which have been evaluated using techniques of structural biology, physico-chemistry, biochemistry and molecular biology. In addition, our design and synthesis of artificial hemoproteins consisted of four-helix bundle are also presented.

2. NADPH OXIDASE

'Professional' phagocytes including neutrophils and macrophages play crucial roles in host defense against pathogens. During phagocytosis, these cells exhibit 'respiratory burst', i.e., intense consumption of O_2 to produce active oxygen species within phagocytic vacuoles for killing and digesting the engulfed microbes (Figure 2). These acitve oxygen species arise from superoxide anions (O_2^-) which are generated at the extracellular surface of vacuolar membranes in response to contact with opsonized microbes (1-4). The respiratory burst is accompanied by oxidation of NADPH and production of O_2^-. The O_2^--producing reaction is catalyzed by a membrane-bound electron-transferring enzyme system, "NADPH oxidase," in which cytochrome $b558$ is the only identified component that contains redox centers.

2.1. Superoxide production by the enzymatic reconstitution system

We demonstrated catalytic production of O_2^- by cytochrome $b558$ purified from porcine neutrophils in an artificial reconstitution system with an exogenous reductase. The isolated cytochrome was resolved into two polypeptides with molecular masses of 60-90 kDa and 19 kDa on sodium dodecyl sulfate-polyacrylamide gel electrophoresis (5). It showed spectroscopic features characteristic of low-spin six coordinated heme (Figure 3). For enzymatic reduction of cytochrome $b558$, we utilized hepatic NADPH-cytochrome P450 reductase purified from rat liver microsomes. Most of the cytochrome in the buffer solution

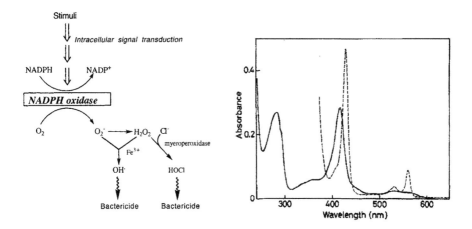

Figure 2 (left):   Production of active oxygens by phagocytes.   Phagocytic cells produce activated oxygen species, including hypochlorous acid (HOCl) and hydroxyl radical (OH•), for killing microorganisms in response to a variety of extracellular stimuli.   The generation of active oxygens occurs cooperatively with the cell morphological changes during phagocytosis (4).

Figure 3 (right):   UV-visible absorption spectra of purified cytochrome b558 in the oxidized (solid line) and reduced (dashed line) forms (5).

was reduced by incubation with the reductase and NADPH under the anaerobic conditions, and was quickly reoxidized by the air (5).   As indicated by measurement of oxygen consumption, the purified cytochrome catalytically reduced oxygen at a rate equal to ~30% of the activity of the phorbol myristate acetate-activated cells on the basis of cytochrome b558 content (Figure 4).   Electron paramagnetic resonance study with a spin trapping agent 5,5-dimethyl-1-pyrroline-1-oxide (DMPO) demonstrated that O_2^- is the exclusive primary product in the reduction of oxygen by the cytochrome (Figure 5).   This gives direct evidence that cytochrome b558 functions as the terminal oxidizing enzyme in the O_2^--generating system of neutrophils.   This also establishes a new functional class of hemoproteins that catalyzes one electron reduction of molecular oxygen.   Purified cytochrome b558 had a midpoint reduction potential (E_m) of -255 mV at pH 7.4 and showed low-spin EPR signals at 10 K (6).   From these results, we concluded that the heme in the six-coordinated low spin state catalyses one electron reduction of O_2.

2.2. Molecular mechanism of the O_2^- production

Molecular oxygen is crucial for oxidative phosphorylation and other important chemical reactions in living cells.   In many processes of the metabolism, O_2 is transformed to reduced or activated forms by heme-containing enzymes.   All known heme-containing enzymes involved in reduction or activation of O_2, such as cytochrome c oxidase or cytochromes P450 have a heme in which the five coordination sites of the iron are occupied by intrinsic ligands and the sixth coordination site is opened for binding O_2 or other extrinsic ligands. The heme of these enzymes forms an iron-O_2 complex during the reduction of O_2, or forms the complexes with respiratory inhibitors. On the other hand, the respiratory burst has been

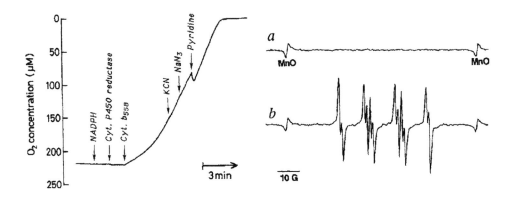

Figure 4 (left): Time course of O_2 consumption by purified cytochrome b558 at 35°C showing no inhibitory effect of cyanide, azide and pyridine. Addition of reagents and proteins to the reaction mixture was indicated by the arrows. The final concentrations were: NADPH, 400 µM; cytochrome P450 reductase, 0.4 µM; purified cytochrome $b558$, 0.4 µM; NaN_3, 2 mM; KCN, 2 mM; pyridine, 0.2 M (6).

Figure 5 (right): Detection of superoxide generated by purified cytochrome $b558$ with DMPO. (a) EPR spectrum of the reaction mixture containing 70 mM DMPO, 400 µM NADPH and 0.1 µM cytochrome P450 reductase in 20 mM HEPES-NaOH (pH 7.4); (b) EPR spectrum after cytochrome $b558$ (0.5 µM) was incubated with the reaction mixture for 2 min (5).

known to be insensitive to the respiratory inhibitors, and cytochrome b558 *in situ* does not form a complex with CO (7). Judging from spectroscopic data, the heme of cytochrome $b558$ is in a low spin six-coordinate state in both the ferric and ferrous forms. Moreover, we confirmed that the respiratory inhibitors gave no effect on the absorption spectra nor on the catalytic activity of purified cytochrome $b558$ in the reconstituted system (Figure 4). These facts indicate strong coordination of both the axial ligands to the heme iron and suggest that cytochrome $b558$ reduces O_2 by a unique mechanism distinct from those of other heme-containing oxidases.

We have analyzed oxidation-reduction kinetics of cytochrome $b558$ purified from porcine neutrophils by stopped-flow and rapid-scanning spectroscopy (8). Reduced cytochrome $b558$ was rapidly reoxidized by O_2, showing spectral changes with clear isosbestic points. The single turnover rate for the reaction with O_2 linearly depended on the O_2 concentration but was not affected by addition of CO. The rate of the reaction decreased with an increase of pH giving a pK_a of 9.7. Under complete anaerobic conditions, ferrous cytochrome $b558$ was oxidized by ferricyanide at a rate faster than by O_2. The thermodynamic analysis showed that the enthalpic energy barriers for the reactions of cytochrome $b558$ are significantly lower than those of the autoxidation of native and modified myoglobins (9) through the formation of the iron-O_2 complex. These findings suggest that the reduction of O_2 by the cytochrome occurs at or near the heme edge by an outer-sphere mechanism. It appears that phagocytic cytochrome $b558$ has evolved the coordination structure of the heme to be specifically adapted for the rapid production of O_2^- even under low concentration of O_2.

In conclusion, we have reconstituted the superoxide-generating enzyme system using cytochrome $b558$ and evidenced that the cytochrome catalyzes one-electron reduction of O_2 to produce O_2^- in the phagocytic NADPH oxidase by the novel enzymatic mechanism (Figure 6).

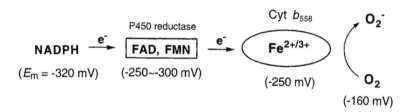

Figure 6: Electron transfer reactions in the reconstituted system with phagocytic cytochrome $b558$.

3. NITRIC OXIDE REDUCTASE

Denitrification is a biological process, in which nitrate (NO_3^-) and/or nitrite (NO_2^-) is converted to N_2 through intermediate formation of nitric oxide (NO) and nitrous oxide (N_2O) ($NO_3^- \rightarrow NO_2^- \rightarrow NO \rightarrow N_2O \rightarrow N_2$). Each process of the denitrification in the bacterial system (prokaryote) is a reduction of nitrogen oxide compounds catalyzed by metalloproteins such as nitrate reductase (NaR; Mo), nitrite reductase (NiR; Cu or Fe), nitric oxide reductase (NOR; Fe) and nitrous oxide reductase (Cu). All enzymes are membrane-bound proteins, and the bacterial processes are coupled with the ATP formation. Therefore, it is generally called as a nitrate respiration.

Recently, Shoun and co-workers (10) found that the denitrification is also present in some fungi (eukaryote). In the fungal dentrification, N_2O is a final product, and NaR (Mo), NiR (Cu) and NOR (Fe) are involved, like in the bacterial system. The fungal NaR and NiR are very homologous to the corresponding enzymes of the bacterial system, suggesting a lateral gene transfer from bacteria to fungi. In addition, fungal NOR is homologous to bacterial cythrome P450s, possibly resulting from the lateral gene transfer (11).

NOR in the denitrification is a chemically unique metallo-enzyme, which catalyzes the N-O bond cleavage and the N-N bond formation from two NO molecules using two electrons and two protons, as follows: $2NO + 2e^- + 2H^+ \rightarrow N_2O + H_2O$. The reaction is catalyzed on the iron site of the enzyme. We have studied relationship of structure and function of NOR using some biochemical and biophysical techniques.

3.1. Fungal nitiric oxide reductase

Fungal nitric oxide reductase (NOR) is a heme-enzyme, that contains a protoheme (Fe-protoporphyrin IX complex) in its active site as a prosthetic group. Its molecular weight is about 46,000. Since its primary structure exhibits a 25% identity in average to those of the heme-monooxygenase cytochrome P450, it belongs to the P450 superfamily, and is usually called P450nor as its custom name. However, it is not a monoxygenase, but catalyzes a NO reduction with a high turnover number (>1,000 sec^{-1}). Since fungal NOR is water-soluble, it possibly relates to the detoxification of NO in fungal cells, but is not coupled with the ATP synthesis. To understand a molecular mechanism of the NO reduction catalyzed by fungal NOR, we have carried out the crystallographic, spectroscopic, kinetic and biochemical studies of fungal NOR.

3.2. Reaction mechanism

The molecular mechanism of the NO reduction by fungal NOR is established as shown in Figure 7A, which has been proposed on the basis of the spectroscopic and kinetic results of the NO reduction by the fungal NOR (12). The ferric enzyme is a resting state, and it combines with NO to yield the ferric NO complex. The ferric-NO complex of the enzyme

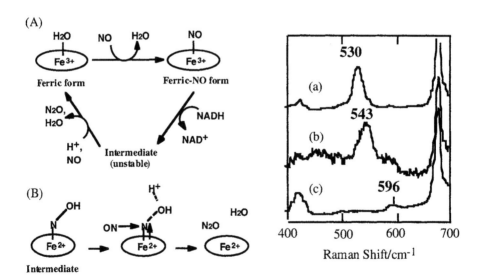

Figure 7A and 7B (left): The proposed reaction mechanism of NO reduction catalyzed by fungal NOR.
Figure 8 (right): Resonance Raman spectra of (a) the ferric-NO, (b) the ferrous-NO complexes, and (c) the reaction intermediate of fungal NOR. The frequency of the Fe-NO stretching is indicated in each state.

is a first and stable intermediate in the NO reduction reaction. The ferric-NO complex is reduced with NAD(P)H without any aid of redox protein mediators, to yield a second and characteristic intermediate. The resultant intermediate *I*, which is transiently generated ($\tau_{1/2}=\sim20\text{sec}$), gives the Soret absorption at 444nm in the optical absorption spectrum. The spectral feature is entirely different from those of the ferric resting state (413nm), the ferric-NO complex (431nm) and the ferrous-NO complex (434nm) of fungal NOR.

To understand the coordination geometry and the electronic structure of *I*, we tried to measure its resonance Raman spectra using the mixed-flow cell (13). The transient Raman spectrum of *I* (Figure 8) showed an isotope sensitive line at 596cm^{-1}, which was shifted to 586cm^{-1} by the ^{15}NO substitution, demonstrating that the line is originated from the Fe-NO stretching mode (νFe-NO). The frequency of the νFe-NO for *I* is higher than those for the ferric-NO (530cm^{-1}) and ferrous-NO (543cm^{-1}) complexes of the enzyme, suggesting that two more electrons donated from NAD(P)H should reside on the Fe-NO moiety, $[Fe^{3+}\text{-NO}]^{2-}$. Since the heme marker lines, $\nu3$ and $\nu4$, of *I* were located at the frequencies observed for the ferrous-NO complex, we could propose the electronic structure of *I* as the $Fe^{2+}\text{-(NO)}^-H^+$ or $Fe^{2+}\text{-NOH}$.

We recently observed the νN-O stretching of the iron-bound NO of *I* at 1330cm^{-1} by the IR spectroscopy (Obayashi, Shiro & Noguchi; unpublished result). The frequency is much lower than those of neutral NO (1840cm^{-1}) and NO$^+$ (2300cm^{-1}), but comparable to that of NO$^-$ (1290cm^{-1}). The result is apparently consistent with the Raman observations, supporting the electronic structure of *I* we proposed.

On the basis of the electronic structure of the second intermediate *I*, we could also propose the mechanism of N$_2$O formation, as shown in Figure 7B, in which another NO molecule attacks to the electron-rich Fe-NO unit in *I*, leading the N-O bond cleavage and the N-N bond formation.

3.3. Molecular structure

We crystallographically determined the molecular structure of the fungal NOR (14). Figure 9 shows the crystal structure of the enzyme in the ferric resting state, which shows a triangular shape with a length of 60Å and a thickness of 30Å. The whole molecule is divided into two domains, the α-helix rich domain and the β-sheet rich domain. The characteristically long I helix is located just above the heme, which is an active site of the enzyme. The basic structural characteristics are quite similar to those of cytochrome P450s, where the crystal structures are available until now. This is well consistent with the fact that the fungal NOR belongs to the P450 superfamily.

The fifth iron coordination site is occupied by a thiolate (S) from Cys352, with the Fe-S bond distance of 2.25Å. The pocket around the sixth iron coordination site, which is constructed by B', F and G helices, is widely opened toward the solvent. Therefore, some external ligands such as CO and NO can coordinate to the sixth iron site with high association rate constants; 2.6×10^7 M^{-1}s^{-1} for NO and 6.1×10^5 M^{-1}s^{-1} for CO (15, 16). The open pocket also allows solvent water molecules come into the active site.

The crystal structure of the ferric-NO complex, which is a first intermediate of fungal NOR in the NO reduction reaction, was also determined (17). This was a first report of the ferric-NO complex of the hemoproteins. The Fe-NO and the N-O bond distances are 1.63Å and 1.16Å, respectively. The NO molecule binds to the iron at the sixth site in slightly tilt ($\angle$S-

Fe-N=160°) and bent (∠Fe-N-O=160°) fashion, in sharp contrast to the highly bent configuration (∠Fe-N-O=110~140°) in the ferrous-NO complexes of hemoproteins.

The determination of the crystal structure of the second intermediate *I* has not yet been successful, due to instability of *I* at room temperature and in solution. We are trying to measure diffraction data of *I* by radiolytical reduction of the ferric-NO complex of fungal NOR in crystal at cryo-temperature.

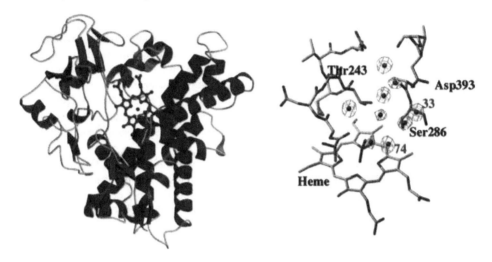

Figure 9 (left): Overall structure of fungal NOR. The α-helices are indicated in green, the β-sheets in red, and the random coils in yellow (17).

Figure 10 (right): Heme environmental structure of fungal NOR showing location of water molecules (red circles) (17).

Proton Delivery pathway: Two protons are utilized in the NO reduction reaction by NOR. The protons are possibly delivered from the solvent water to the active site through hydrogen-bonding networks. To characterize the proton delivery pathway, we examined the hydrogen-bonding network in the crystal structure of the ferric-NO complex of fungal NOR. As was illustrated in Figure 10, a hydrogen-bonding network which connect the active site to the solvent region is identified; that is Wat74--Ser286--Wat33--Asp393--solvent. The Wat74 is located 3.10Å from the iron-bound NO, and is the closest for the water molecules. Since Ser286 is a key residue in this hydrogen-bonding network, we prepared some mutants of fungal NOR, in which Ser286 was replaced with Val or Thr (17, 18). The Ser286Val and Ser286Thr mutants exhibited no or little activity in the NO reduction reaction. In the crystal structures of the mutants, Wat74 is removed from its original position, and therefore the connection between the active site and the solvent region was completely disrupted. Thus, we concluded that the hydrogen-bonding network, Wat74--Se286--Wat33--Asp393--solvent, acts as the proton delivery pathway in the NO reduction reaction by fungal NOR.

Interaction with Electron Donor: The NO reduction reaction catalyzed by fungal NOR does not require any redox proteins (10). This facts allows us to consider that the electron donor, NAD(P)H, should directly bind to the enzyme, and should directly donate two electrons to the iron site. Indeed, fungal NORs discriminate between NADH and NADPH. However, a classical NAD(P)H binding motif is not found in the fungal NOR structure. Unfortunately, we have not yet obtained the single crystal of the enzyme-NAD(P)H complex by co-crystallization and NAD(P)H soaking techniques. In the structure of fungal NOR, it was found that a cluster consisting of the positively charged residues (Lys62, Arg64, Lys291, Arg392) is present near the sixth iron site beneath the B' helix (18). Replacement of the residue in the cluster by the point-mutation retarded the enzymatic activity of fungal NOR, suggesting that the positively charged cluster might be an interaction site with NAD(P)H.

Our structural and functional studies on fungal NOR could provide us implications on the NO reduction mechanism by NORs of bacterial denitrification system, which are more widely distributed. In the fungal system, two electrons are injected into one NO molecule at the single iron site. On the other hand, it is likely that the active site of bacterial NOR might be designed for two reactions of one-electron injection to one NO molecules at the bi-nuclear iron site, followed by disproportionation of resultant two NO^- ($2NO + 2e^- + 2H^+ \rightarrow 2NO^- + 2H^+ \rightarrow N_2O + H_2O$). This hypothetical mechanism should be examined when a crystal structure of bacterial NOR is determined.

4. OXYGEN SENSOR PROTEIN FixL

Molecular oxygen is the most pivotal of the environmental factors for living organisms. Cells prefer to oxygen in various living processes; energy production in oxidative phosphorylation, metabolite production by oxidation, and generation of reactive oxygen species (O_2^-, NO) for biodefense. Cellular adaptation to changes in oxygen tension, usually to hypoxia, is widely observed from bacteria to mammals (Figure 11) (19). Acute responses such as ventilation of blood flow do not require *de novo* protein synthesis whereas non-acute

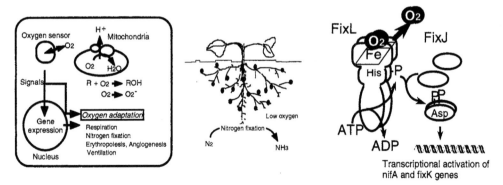

Figure 11 (left): Oxygen adaptation of living organisms. Most oxygen adaptations are conducted at the transcriptional level.
Figure 12: A (middle), Rhizobial bacteria fix nitrogen gas to ammonia in the anaerobicaly symbiosis within plant root nodules. B (right), Two-component signal tranduction of FixL and FixJ directs expression of the nitrogen fixation-related genes at low oxygen tension.

or chronic responses such as erythropoiesis undergo changes in gene expression (20). Heme, flavin and non-heme iron proteins are possible oxygen sensors in the signaling cascade(s) although investigations on their natures have been limited (21-23).

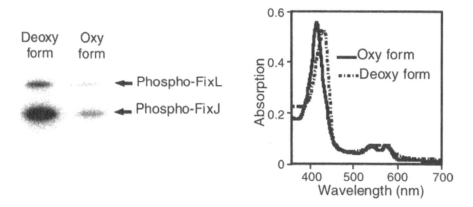

Figure 13 (left): In vitro phosphorylation of FixL and FixJ. Oxygen-free (deoxy-) FixL catalyzes phosphoryl transfer from ATP to FixJ.
Figure 14 (right): FixL exhibits the reversible spectral change upon association/dissociation of oxygen.

Rhizobial FixL/FixJ system is a paradigm of biological oxygen sensors, which senses low oxygen tensions to regulate expression of the genes involved in nitrogen fixation in the symbiotic anaerobic state within the plant root nodules (Figure 12) (24). FixL and FixJ have also been known to be a pair of the sensory histidine kinase and the cognate response regulator of the bacterial two-component regulatory systems (25). FixL autophosphorylates a conserved histidine residue and transfers the phoshoryl group to FixJ at low oxygen tensions to express the fixK and nifA genes whereas FixL represses the kinase activity under the aerobic conditions (Figure 13). In other words, the sensor domain contains a heme and the dissociation of O_2 from the heme moiety triggers autophosphorylation in the kinase domain at low oxygen tension.

FixL derived from *Rhizobium meliloti* (RmFixL) consists of a membrane-anchoring domain, a sensor domain and a histidine kinase domain. The histidine kinase domain is further divided into an autophosphorylation subdomain containing a phospho-accepting His285 (H box) and a catalytic subdomain containing an ATP-binding site (N box and G box) (25).

Histidine kinases in the two-component regulatory systems share the well-conserved kinase domains, and mechanism of the intra- and inter-molecular signaling following the initial ligand-binding may be common although the sensor domains are individually different. Thus, FixL/FixJ is the most useful model system of the ubiquitous two-component regulatory systems because oxygen-free form (kinase-active) and oxygen-bound form (kinase-inactive) of FixL are spectroscopically distinguishable (Figure 14) (26).

In this section, we will review the molecular architecture of the FixL/FixJ system based on our recent studies.   We do not hesitate to refer to other two-component systems to help one to understand the signaling mechanisms of FixL/FixJ system.

4.1. N-terminal membrane-anchoring domain of FixL

Since the sensory histidine kinases sense the extracellular ligands, the sensor domians of many kinases are localized in the cytoplasmic membranes (25).   Like other sensory kinases, FixL derived from *Rhizobium meliloti* contains the hydrophobic membrane-spanning segments at the N-terminus (27).   However, the localization of FixL in the membranes seems non-essential because molecular oxygen is membrane-permeable and oxygen-binding site (heme) is exposed to the cytoplasm.   In fact, *Bradyrhizobium japonicum* FixL possesses the hydrophilic fragment at the N-terminus (28).   It is also revealed that a deletion of the membrane-spanning regions from Rm FixL results in no significant defects on the signaling functions *in vivo* and *in vitro* (Nakamura et al., in preparation).

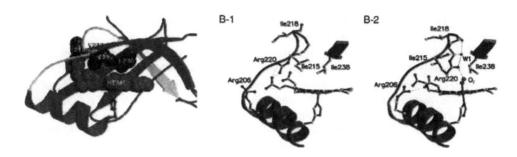

Figure 15: Crystal structures of the sensor domains of FixL. A, Rm FixL sensor domain (33). B, Heme vicinity of Bj FixL sensor domain (35).

4.2. Oxygen-binding domain of FixL

FixL contains a protoheme as an oxygen-binding site.   Optical absorption spectra of oxygen-bound (oxy-) form and ferrous oxygen-free (deoxy-) form are quite similar to hemoglobin. Spin state of the heme iron in FixL seems to determine the kinase activity; that is, high spin- or deoxy-FixL has elevated the kinase activity while the low spin- or oxy-form is inactive (29).   Local structural changes of the heme and its vicinity caused by O_2 association/dissociation has been supposed to initiate the intramolecular regulation of the kinase domain by analogy of the R-T state conversion (29-31).   Recently, crystallographic structures of the ferric high spin- and low spin heme domains derived from *B. japonicum* have revealed that flattering/doming of the heme plane by ligand binding causes changes of the hydrogen bonds and the salt bridges between the heme propionates and the "FG regulatory loop" in the distal pocket of the heme (32).   Based on these observations, they have proposed that the conformational change of the loop initially transduces the ligand-binding signal to the kinase domain. However, we found that CO-bound FixL, a low spin form, significantly autophosphorylated and transferred the phosphoryl group to FixJ (unpublished observation). This result gave rise to an argument against the correlation between the kinase activity and the spin state accompanying the heme doming. Since the amino acids in the "FG regulatory loop" play the critical roles on the kinase regulation (Figure 15A) (33, 34), we have proposed that the bound oxygen directly interacts with the amino acids of the distal G helix to inactivate the kinase acitvity. It should be noted that their folds are distinct from those of serine/threonine revealed that oxygen-bound sensor domain from *B. japonicum* contains the hydorgen bond between the bound oxygen and conserved arginine at the G helix, which agrees with our idea (Figure 15B) (35).

4.3. Autophosphorylation domain and catalytic domain of FixL

Since phosphoryl transfer to FixJ and dephosphorylation from FixJ are oxygen-independent, regulatory mechanisms coupled with ligand-binding are restricted to the autophosphorylation reaction (36). Therefore, regulatory mechanism between the sensor and the kinase domains is a crucial issue to be elucidated. Autophosphorylation domains (H box) and catalytic domains (N and G boxes) are much conserved among the sensory histidine kinases (25). While these domains of FixL have not been studied enough, functional and structural analyses are available from the osmosensor EnvZ protein.

The H box is responsible to the dimerization of the kinases by forming a pair of the two helical bundles (37). Two conserved histidines are exposed outside in the opposite directions each other. This configuration enables the histidines to accept the phosphoryl group from each catalytic domain and tranfer it to the response regulator. Catalytic domain contains the conserved asparagine and glycines to stabilize the bound ATP, and the overall fold is found in ATPases (DNA gyrase and Hsp90) to form a superfamily of ATP-driven enzymes (38). It should be noted that their folds are distinct from those of serine/threonine or tyrosine kinases of other signal transduction familes.

ATP is hydrolysed at the catalytic domain of one monomer, and the phosphoryl group is transferred to the histidine of the other monomer (trans-phosphorylation) (39, 40; Nakamura , unpublished observation). However, it remains to be elucidated whether the sensor domains act on the H box or the catalytic domain. It is also unknown which sensor domain regulates the autophosphorylation of one monomer.

4.4. FixJ as a transcriptional regulator

FixJ is phosphorylated at Asp54 of the N-terminal receiver domain. Phospho-FixJ is dimerized by the interaction with the individual receiver domains (41). The C-terminal half contains a helix-turn-helix motif like as other transcriptional factors, and is activated to bind to the fixK and nifA promoters by phosphorylation (Tamura, unpublished observation). However, it is unknown whether or not the dimerization is directly required for DNA binding activity.

It has passed only 15 years since involvement of protein phosphorylations in bacterial adaptive responses was first discovered (42), and one can recognize in the last decade of the 20th century that the two component signal transduction systems are widely spread in prokaryotic and eukaryotic organisms. As mentioned above, little of the regulatory mechanisms have been revealed at molecular or atomic level. We believe that knowledge obtained from the FixL/FixJ system will shed novel and profound insights into the ubiquitous signal transducing machinery.

5. DESIGN AND SYNTHESIS OF ARTIFICIAL HEMOPROTEINS

Design of novel proteins, in which simple peptides are synthesized to assemble into well defined structural motifs of native proteins, are providing great insights into the basis of protein stability, activity, and the folding mechanisms. One of these approaches is making molecular maquettes with incorporated cofactors, which are simplified versions of the native metalloproteins involved in redox catalysis and in respiratory and photosynthetic energy conversion. Robertson et al. have synthesized stable and geometrically well-defined four-helix bundle proteins from a helix-loop-helix peptide that self-assembles into the four-helix dimer with ligation of up to four molecules of protoheme or heme B (43). They exhibited strikingly native-like spectroscopic and electrochemical properties, i.e., the optical absorption and electron-spin resonance spectra characteristic of low-spin b-type cytochromes with bis-histidine ligation, and midpoint redox potentials between -80 and -230 mV with heme-heme redox interaction. The synthetic hemoprotein as a maquette for cytochrome bc_1 complex has been modified by covalent attachment of a porphyrin dimer for a photosynthetic reaction center maquette (44), and an iron sulfur cluster for a ferredoxin maquette (45). These approaches lead to deeper understanding of how their counterpart native proteins form the specific redox centers and how they function in complicated biological systems.

5.1. Complex of four-helix bundle and heme A
In this contribution, we have introduced heme A, a prosthetic group of cytochrome c oxidase, into a designed four-helix bundle to extend the synthetic system to a cytochrome c oxidase maquette (46). We used a synthetic peptide with the sequence; Ac-CGGGE LWKLH EELLK KFEEL LKLAE ERLKK L-CONH$_2$, here called α, for ligation of the heme. After the disulfide-bond formation by oxidation of the N-terminal Cys thiol, the resulting 62-residue, di-α-helical peptide (α_2) assembles into a four-helix bundle to form two bis-histidyl heme binding sites at positions 10, 10' in [α_2]$_2$. Heme A differs from heme B in structure; a formyl group and a C15 isoprenyl side chain replace one of the methyl groups and one of the vinyl groups, respectively (Figure 1). Thus, both the steric and electrostatic effects of these substitutions can be expected on properties of the synthetic protein with heme A. The synthetic protein binds two heme A molecules in the four-helix unit. The heme A-incorporated protein exhibits well-defined absorption spectra in both the ferric and ferrous states, and an electron paramagnetic resonance spectrum characteristic of a low spin heme in the ferric form. The midpoint redox potential of the bound heme A was single and determined to be -45 mV (vs. S.H.E.) at pH 8.0, which is higher than those of the bound protoheme and of histidine-heme A hemochrome, and lower than that of the natural counterpart, heme a of cytochrome c oxidase. The observation of a single midpoint redox potential for [heme A-α_2]$_2$ and a pair of midpoints for [heme B-α_2]$_2$ indicates that the di-α-helical monomers are oriented in an *anti* topology (disulfides on opposite sides of bundle) in the former (lacking heme-heme electrostatic interaction) and *syn* in the latter. Self-

assembly of the mixed cofactor {heme A/heme B-α_2} was accomplished by addition of a single equivalent of each heme A and heme B to [α_2]$_2$. The single midpoint redox potential of heme B, E_{m8} = -200 mV, together with the split midpoint redox potential of heme A in {[heme A/heme B-α_2]$_2$}, E_{m8} = +28 mV (33%) and –65 mV (67%), indicated the existence of both *syn* and *anti* topologies of the two di-α-helical monomers in this four helix bundle (Figure 16). The results indicate that heme peripheral structure controls the orientation of the di-α-helical monomers in the four-helix bundle which are interchangeable between *syn* and *anti* topologies.

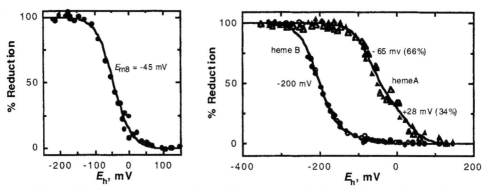

Figure 16. Potentiometric titration of [heme A-α_2]$_2$ (left) and [heme A/heme B-α_2]$_2$ (right). The reduced fraction of the bound hemes are plotted against ambient redox potential, which is indicated against standard hydrogen electrode (S.H.E). The Nernst curves were fitted to the data with midpoint redox potentials indicated in the plots and the *n* number of 1.

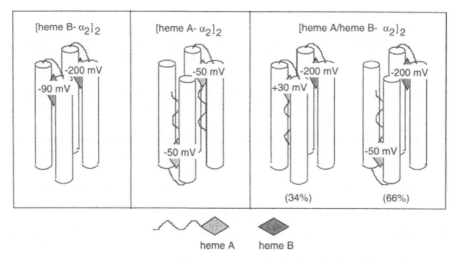

Figure 17: Global topologies of heme A-, heme B-binding [α_2]$_2$ and heteroheme-binding [α_2]$_2$

In the reduced form, [heme A-α_2]$_2$, reacts quantitatively to form [CO-heme A-α_2]$_2$ as evidenced by optical spectroscopy. The synthetic [heme A-α_2]$_2$ can be enzymatically reduced by NAD(P)H with natural reductases under anaerobic conditions, and reversibly oxidized by O_2 to the ferric form, suggesting a possibility of *de novo* creation of redox enzymes (46). We are now developing a general method to design artificial amino acid sequences folding into a desired tertiary structure, which is applicable not only to simple symmetrical structures such as a helical bundle but to complex asymmetrical structures (47, 48).

REFERENCES

1. A. R. Cross and O. T.G. Jones, Biochim. Biophys. Acta, 1057 (1991) 281.
2. F. Morel, J. Doussiere and P. V. Vignais, Eur. J. Biochem., 201 (1991) 523.
3. A. W. Segal and A. Abo, Trends. Biochem. Sci., 83 (1993) 1785.
4. Y. Isogai, T. Tsuyama, H. Osada, T. Iizuka and K. Tanaka, FEBS Lett., 380 (1996) 263.
5. Y. Isogai, Y. Shiro, A. Nasuda-Kouyama, and T. Iizuka, J. Biol. Chem., 266 (1991) 13481.
6. Y. Isogai, T. Iizuka, R. Makino, T. Iyanagi and Y. Orii, J. Biol. Chem., 268 (1993) 4025.
7. T. Iizuka, S. Kanegasaki, R. Makino, T. Tanaka and Y. Ishimura, J. Biol. Chem., 260 (1985) 12049.
8. Y. Isogai, T. Iizuka, and Y. Shiro, J. Biol. Chem., 270 (1995) 7853.
9. Y. Shiro, T. Iwata, R. Makino, M. Fujii, Y. Isogai and T. Iizuka, J. Biol. Chem., 268 (1993) 19983.
10. K. Nakahara, T. Tanimoto, K. Hatano, K. Usuda and H. Shoun, J. Biol. Chem., 268 (1993) 8350-8355
11. H. Kizawa, D. Tomura, M. Oda, A. Fukamizu, T. Hoshino, O. Gotoh, T. Yasui and H. Shoun, J. Biol. Chem., 266 (1991) 10632.
12. Y. Shiro, M. Fujii, T. Iizuka, S. Adachi, K. Tsukamoto, K. Nakahara and H. Shoun, J. Biol. Chem., 270 (1995) 1617.
13. E. Obayashi, S. Takahasi and Y. Shiro, J. Am. Chem. Soc., 120 (1998) 12964.
14. S.-Y. Park, H. Shimizu, S. Adachi, A. Nakagawa, I. Tanaka, K. Nakahara, H. Shoun, H. Nakamura, T. Iizuka and Y. Shiro, Nat. Struct. Biol., 4 (1997) 827.
15. Y. Shiro, M. Fujii, M. Kato, T. Iizuka, K. Nakahara and H. Shoun, Biochemistry, 33 (1994) 8673.
16. E. Obayashi, K. Tsukamoto, S. Adachi, S. Takahashi, M. Nomura, T. Iizuka, H. Shoun and Y. Shiro, J. Am. Chem. Soc., 119 (1997) 7807.
17. H. Shimizu, E. Obayashi, Y. Gomi, H. Arakawa, S.-Y. Park, H. Nakamura, S. Adachi, H. Shoun and Y. Shiro, J. Biol. Chem., 275 (2000) 4816.
18. H. Shimizu, S.-Y. Park, D.-S. Lee, H. Shoun and Y. Shiro, J. Inorg. Biochem., 81 (2000) 191.
19. H. F. Bunn and R. O. Poyton, Physiol Rev.,76 (1996) 839.
20. G. L. Semenza, Cell, 98 (1999) 281.
21. M. A. Goldberg, S. P. Dunning and H. F. Bunn, Science, 242 (1988) 1412.
22. J. M. Gleadle, B. L. Ebert and P. J. Ratcliffe, Eur. J. Biochem., 234 (1995) 92.
23. D. Wang, C. Youngson, V. Wong, H. Yeger, M. C. Dinauer, E. Vega-Saenz de Miera, B. Rudy and E. Cutz, Pro. Natl. Acad. Sci. USA, 93 (1996) 13182.

24. H.-M.Fischer, Microbiol. Rev., 58 (1994) 352.

25. J. B. Stock, A. J. Ninfa and A. M. Stock, Microbiol. Rev., 53 (1989) 450.

26. M. A. Gilles-Gonzalez, G. S. Ditta and D. R. Helinski, Nature, 350 (1991) 170.

27. A. F. Lois, G. S. Ditta and D. R. Helinski, J. Bacteriol., 175 (1993) 1103.

28. M. A. Gilles-Gonzalez, G. Gonzalez, M. F. Perutz, L. Kiger, M. C. Marden and C. P oyart, Biochemistry, 33 (1994) 8067.

29. M. A. Gilles-Gonzalez, G. Gonzalez and M. F. Perutz, Biochemistry, 34 (1995), 232.

30. K. Tamura, H. Nakamura, Y. Tanaka, S. Oue, K. Tsukamoto, M. Nomura, T. Tsuchiya, S. Adachi, S. Takahashi, T. Iizuka and Y. Shiro, J. Am. Chem. Soc., 118 (1996) 9434.

31. H. Miyatake, M. Mukai, S. Adachi, H. Nakamura, K. Tamura, T. Iizuka, Y. Shiro, R. W. Strange and S. S. Hasnain, J. Biol. Chem., 274 (1999) 23176.

32. W. Gong, B. Hao, S. S. Mansy, G. Gonzalez, M. A. Gilles-Gonzalez and M. K. Chan, Pro. Natl. Acad. Sci. USA, 95 (1998) 15177.

33. H. Miyatake, M. Mukai, S.-Y. Park, S. Adachi, K. Tamura, H. Nakamura, K. Nakamura, T. Tsuchiya, T. Iizuka and Y. Shiro, J. Mol. Biol., 301 (2000) 415.

34. M. Mukai, K. Nakamura, H. Nakamura, T. Iizuka and Y. Shiro, Biochemistry, 39 (2000) 13810.

35. W. Gong, B. Hao and M. K. Chan, Biochemistry, 39 (2000) 3955.

36. M. A. Gilles-Gonzalez, and G. Gonzalez, J. Biol. Chem., 268 (1993) 16293.

37. C. Tomomori, T. Tanaka, R. Dutta, H. Park, S. K. Saha, Y. Zhu, R. Ishima, D. Liu, K. I. Tong, H. Kurokawa, H. Qian, M. Inouye and M. Ikura, Nat. Struct. Biol., 6 (1999), 729.

38. T. Tanaka, S. K. Saha, C. Tomomori, R. Ishima, D. Liu, K. I. Tong, H. Park, R. Dutta, L. Qin, M. B. SwindellS, T. Yamazaki, A. M. Ono, M. Kainosho, M. Inouye and M. Ikura, Nature, 396 (1998) 88.

39. Y. Yang and M. Inouye, Pro. Natl. Acad. Sci. USA, 88 (1991) 11057.

40. E. G. Ninfa, M. R. Atkinson, E. S. Kamberov and A. J. Ninfa, J. Bacteriol., 175 (1993) 7024.

41. S. Da Re, J. Schumacher, P. Rousseau, J. Fourment, C. Ebel and D. Kahn, Mol. Microbiol., 34 (1999) 504.

42. A. J. Ninfa, and B. Magasanik, Pro. Natl. Acad. Sci. USA, 83 (1986) 5909.

43. D. E. Robertson, R. S. Farid, C. C. Moser, S. E. Mulholland, R. Pidikit, J. D. Lear, A. J. Wand, W. F. DeGrado and P. L. Dutton, Nature, 368 (1994) 425.

44. F. Rabanal, W. F. Degrado and P. L. Dutton, J. Am. Chem. Soc., 118 (1996) 473.

45. B. R. Gibney, S. E. Mulholland, F. Rananal and P. L. Dutton, Proc. Natl. Acad. Sci. USA, 93 (1996) 15041.

46. B. R. Gibney, Y. Isogai, F. Rabanal, K. S. Reddy, A. M. Grosset, C. C. Moser and P. L. Dutton, Biochemistry, 39 (2000) 11041.

47. Y. Isogai, M. Ota, T. Fujisawa, H. Izuno, M. Mukai, H. Nakamura, T. Iizuka and K. Nishikawa, Biochemistry, 38 (1999) 7431.

48. Y. Isogai, A. Ishii, M. Ota and K. Nishikawa, Biochemistry, 39 (2000) 5683.

Molecular Anatomy of Cellular Systems
I. Endo et al., (editors)
© 2002 Elsevier Science B.V. All rights reserved.

Unity and diversity in biological oxidation

Johannis A. Duine

Research Institute for Bioresources, Okayama University, 2-20-1 Chuo, Kurashiki, 710 0046 Japan

Half a century ago, from insight obtained in metabolic pathways it was concluded that biological oxidation concerns the transfer of reducing equivalents from a donor to an acceptor, implicating that unity exists with respect to the basic mechanism of it. On the other hand, research in the past twenty years, especially that on enzymes having a quinone as cofactor, has revealed several new oxidoreductases (quinoproteins) that catalyse an oxidation reaction that is also carried out by e.g. a flavoprotein. This review attempts to give insight into such diversity by distinguishing the enzymes involved in oxidation of alcohols, and (unphosphorylated) sugars, according to the type of cofactor or coenzyme they use for this. In addition, examples are given showing that the diversity improves the possibilities for enzyme application.

For those who have been investigated, it appears that the "new" oxidoreductases do not distinguish themselves from the "old" ones with respect to their mechanism, direct hydride transfer being the dominating principle for transfer of reducing equivalents from alcohols and sugars. Thus, why do so many enzymes, quite different in identity of cofactor/coenzyme as well as primary and tertiary structure, catalyse the oxidation of a certain substrate with essentially the same chemical mechanism. In trying to answer this question, two possibilities are discussed. In the first one it is assumed that the diversity has no meaning, the different enzymes regarded as remaining relics from the past. In the other one the opposite is proposed, and the difficulty to explain the meaning ascribed to our ignorance about adequate functioning of an enzyme in its specific context. Since from our present knowledge it appears impossible to decide which explanation is correct, the conclusion is that enzymology should focus not only on mechanistic details of the *in vitro* substrate conversion step but also on enzyme functioning related to the demands imposed by a particular organism or environment.

1.INTRODUCTION

As formulated already more than half a century ago, unity exists with respect to the mechanism of biological oxidation reactions as it always concerns the transfer of reducing equivalents from a donor to an acceptor. This applies not only to the reactions proceeding in common metabolic pathways like the citric acid cycle or glycolysis but also to those in the astonishing variety of microbial oxidation pathways. From present day basic principles of enzymology, it is understandable that efficient catalysis requires that "each substrate has its own enzyme". Although firm data substantiating it are lacking, it is also felt that optimisation of a certain enzyme with respect to the task to be fulfilled in the specific organisms, results in variation on the theme, i.e. in structural variants of the enzyme (orthologues). However, to explain that oxidation of a certain substrate can be carried out by quite different enzymes with respect to structure and cofactor, requires more from our imagination. Research in the past twenty years, especially that on the new quinone cofactors (Fig. 1), has thrown up many examples for diversity in enzymes catalysing the same reaction, sometimes even the oxidation of the same substrate. To illustrate the diversity, enzymes oxidizing alcohols or sugars are rubricated according to the cofactor and coenzyme they need for this. In a number of cases it will be shown how enzyme application benefits from diversity with examples related to e.g. kinetic resolution of racemates or a role as sensor in analytical devices. Finally, two possibilities are discussed to explain the existence of "diversity in unity".

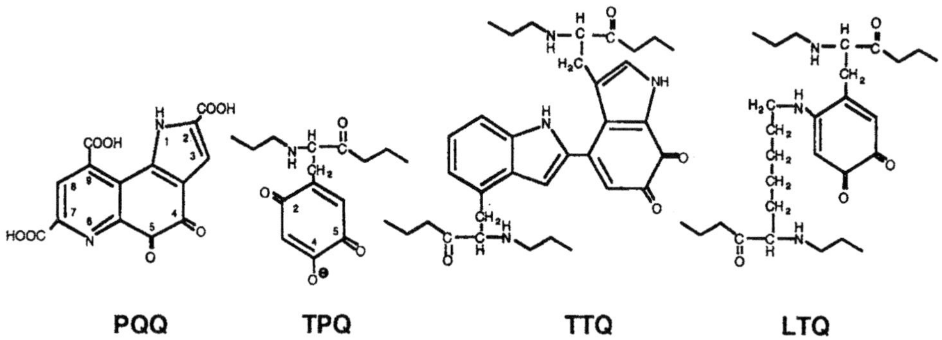

Fig. 1. Structures of quinone cofactors: PQQ, pyrroloquinoline quinone; TPQ, topaquinone; TTQ, tryptophyl-tryptophanquinone; LTQ, lysyl-topaquinone.

1.1. On cofactors, coenzymes, prosthetic groups, dehydrogenases, and oxidases

Especially in oxidative reactions, the amino acid residues in a protein are as such unable to catalyse the reaction: they need the assistance of a cofactor, coenzyme or prosthetic group. It should be noted that compounds that affect enzyme activity but do not directly participate in the reaction (e.g. allosteric modulators or activators) are excluded by this definition.

A coenzyme is a cosubstrate transiently associated with the enzyme. After the reaction it dissociates and is regenerated by another enzyme. This situation is illustrated by NAD, acting as coenzyme for NAD-dependent alcohol dehydrogenase (ADH):

$$\text{ADH}$$

$$NAD^+ + \text{Ethanol} \rightarrow H^+ + NADH + \text{Acetaldehyde}$$

As a consequence of the mechanism, the reaction is affected by the status of the redox balance in the organism (the ratio of NAD^+ to NADH in the cytosol). NAD^+ functions as coenzyme for several different NAD-dependent dehydrogenases, NADH as coenzyme for several NADH-dependent reductases, and as substrate for NADH-dehydrogenase which forms part of the respiratory chain. Since the NAD(P)-dependent ADHs catalyse kinetically reversible reactions, all these factors affect the direction and rate of their reactions.

In contrast to coenzymes, cofactors remain bound to the enzyme during the whole reaction cycle so that regeneration of the cofactor should take place on the enzyme itself. This is illustrated by pyrroloquinoline quinone, PQQ (Fig. 1), acting as cofactor in the reaction of methanol dehydrogenase (MDH) with methanol, and being oxidized in the enzyme with a special component of the respiratory chain, cytochrome c_L, after that:

$$E\sim PQQ + Methanol \rightarrow E\sim PQQH_2 + Formaldehyde$$

$$E\sim PQQH_2 + 2 \text{ ferricytochrome } c_L \rightarrow E\sim PQQ + 2 H^+ + 2 \text{ ferrocytochrome } c_L$$

To distinguish it from an NAD-dependent dehydrogenase, a dehydrogenase like MDH is sometimes called a "dye-linked dehydrogenase". The reason is that although such a type of dehydrogenase is always coupled to a component of the respiratory chain (but see nicotinoproteins for exceptions), in the absence of this natural electron acceptor an artificial one such as a dye can be used. It will be clear from this that the activity of a dye-linked dehydrogenase in the cell is not directly affected by the redox balance (NAD/NADH ratio) but by the redox status of the respiratory chain.

A prosthetic group is a cofactor that is covalently bound to the protein. An example has already been presented above, cytochrome c, in which the heme c is attached to the protein chain. In this review, no distinction will be made between prosthetic group and cofactor.

An oxidase uses O_2 as electron acceptor, the reoxidation of the cofactor producing H_2O_2 (hydrogen peroxide). The example shown concerns glucose oxidase, a flavoprotein which has FAD as cofactor:

$$E\sim FAD + D\text{-Glucose} \rightarrow E\sim FADH_2 + D\text{-Gluconolactone}$$

$$E\sim FADH_2 + O_2 \rightarrow E\sim FAD + H_2O_2$$

Oxidases are either extra-cellular or peroxisomal enzymes. In some cases, the formed H_2O_2 is used by an H_2O_2-consuming enzymatic reaction, in others it is destroyed by catalase. From an energetic point of view, oxidases are uneconomical since the substrate conversion step does not contribute to useful energy production, the necessity to get rid of H_2O_2 even costing energy for the biosynthesis of catalase.

2. ALCOHOL OXIDATION

2.1. Introduction

The ability to oxidize alcohols is widespread among living organisms. In animals and plants, the reaction proceeds *via* an NAD-dependent alcohol dehydrogenase. Based on amino acid sequence comparison, three families can be distinguished: the short chain, metal independent family; the long chain, iron-activated family to which ADH II from *Zymomonas mobilis* belongs; the medium chain, zinc-containing family. The latter consists of six classes: e.g. Class I, encompassing the well-known enzyme from liver, occurring in numerous organism- and organ-related forms (orthologues and paralogues, resp.); e.g. Class III (which oxidizes higher alcohols but also functions as glutathione-dependent formaldehyde dehydrogenase). However, the enzymes in all these classes use NAD(P) as coenzyme and are structurally related, a form of Class III most probably being the ancestor of this large family of enzymes.

Many microbes can utilize alcohols as sole carbon and energy source, either in axenic or in mixed culture. Although ethanol is the most universal alcohol used, the range seems unlimited since certain specialized microbes utilize the smallest one, methanol, others a large one like the xenobiotic polyethylene glycol (PEG) 20.000. Besides direct, also indirect utilization occurs in pathways in which alcohols are an intermediate, e.g. in the degradation of alkanes. As illustrated by the summarizing Table 3, microbial oxidation of an alcohol to the corresponding aldehyde is carried out by quite different enzymes, not only for alcohols in general but also for an individual alcohol like ethanol or methanol.

2.2. NAD(P)-dependent alcohol dehydrogenases

Several examples ca n be found where alcohol oxidation by a microbe is carried out by an NAD(P)-dependent alcohol dehydrogenase belonging to one of the three families indicated above. Only one NAD(P)-dependent methanol dehydrogenase has so far been found, that from *Bacillus methanolicus* [1]. However, the enzyme itself is in fact a nicotinoprotein (see below), the main part of its NAD-dependency based on the action of an activator protein. To oxidize formaldehyde, ADH class III from Actinomycetes (a group of Gram-positive bacteria) requires mycothiol (1-D-myo-inosityl-2-(N-acetylcysteine)amido-2- deoxy-a-D-glucopyranoside) instead of glutathione (Fig. 2) for the formation of the active substrate adduct [2]. Despite this coenzyme preference, the enzyme shows significant sequence similarity with its glutathione-using counterparts [3].

ADHs have been found in Archae which use factor F420 (a deazaflavin) as coenzyme instead of NAD(P) [4].

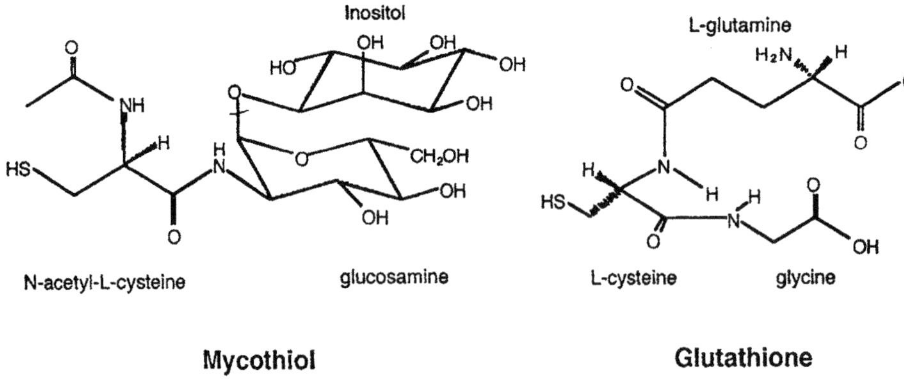

Mycothiol   **Glutathione**

Fig. 2. Structure of mycothiol and glutathione.

2.3. Nicotinoprotein alcohol dehydrogenases

A nicotinoprotein is an enzyme that uses NAD(P) as cofactor instead as coenzyme. According to the definition of a cofactor, NAD(P) should remain bound during the catalytic cycle of a nicotinoprotein. Since also no exchange of reducing equivalents with externally added coenzyme occurs, as a consequence the enzymes cannot be assayed by following NADH formation when substrate and NAD is added. However, NDMA (4-nitroso-NN'-dimethylaniline) is a good artificial electron acceptor for most of them. Only in one case, methanol:NDMA oxidoreductase (MNO)from the bacterium *Amycolatopsis methanolica*, it has been found that the nicotinoprotein enzyme is coupled to the respiratory chain [5]. In other cases, it is likely that the nicotinoprotein catalyzes interconversions, e.g. in the case of formaldehyde dismutase (FDM) from a *Pseudomonas* strain [6] and nicotinoprotein alcohol dehydrogenase (np-ADH) from *Rhodococcus erythropolis* [7]:

FDM

$$\text{Formaldehyde} + \text{E} \sim \text{NAD} \rightarrow \text{Formate} + \text{E} \sim \text{NADH}$$

$$\text{Formaldehyde} + \text{E} \sim \text{NADH} \rightarrow \text{Methanol} + \text{E} \sim \text{NAD}$$

np-ADH

$$\text{Alcohol}_1 + \text{E} \sim \text{NAD} \rightarrow \text{E} \sim \text{NADH} + \text{Aldehyde}_1$$

$$\text{Aldehyde}_2 + \text{E} \sim \text{NADH} \rightarrow \text{E} \sim \text{NAD} + \text{Alcohol}_2$$

2.4. Flavoprotein alcohol dehydrogenases and oxidases

Characterization of polyethylene glycol dehydrogenase (PEG-DH) from Sphingomonads bacteria has revealed a new type of alcohol-oxidizing enzyme, membrane-bound, flavoprotein (FAD-containing) alcohol dehydrogenase [8]. Enzymes with high sequence similarity and the FAD signature have been found (Table 2) for octanol [9], and 4-nitrobenzyl alcohol [10]. A similar enzyme is involved in the oxidation of the sugar L-sorbose by *Gluconobacter oxydans* [11] and perhaps in that in oxidation of the polyol sorbitol [12]. Although the flavoprotein alcohol dehydrogenases form a distinct group, they are structurally related to a group of flavoprotein alcohol oxidases consisting of methanol oxidase [13]and aryl alcohol oxidase [14], enzymes produced by yeasts and fungi, respectively. Vanillyl alcohol oxidase is produced also by a fungus but it is structurally related to another group of flavoproteins that have FAD as covalently bound cofactor [15]. Also hexose oxidase (see below) belongs to the latter group.

Table 1

Flavoprotein (FAD-containing) alcohol dehydrogenases

ENZYME	ORGANISM	ENZYME	ORGANISM
Octanol DH	*Ps. oleovorans*	Polyethylene glycol DH	*Sphingomonads*
Choline DH	*E. coli* and others	4-Nitrobenzyl alc. DH	*Pseudomonas* sp.

2.5. Quinoprotein alcohol dehydrogenases

Methanol dehydrogenase, occurring in Gram-negative bacteria utilizing methanol or methane as carbon and energy source, was the first enzyme in which the cofactor PQQ was found. At that time, production of single cell protein, consisting of methanol-grown bacterial cell mass seemed economically feasible. As has been described in the "PQQ story"[16], the relatively large amount of PQQ needed for structure elucidation was isolated from that material. Subsequently, several other PQQ-containing alcohol dehydrogenases were discovered for ethanol [17, 18], quinate [19], glycerol, and sorbitol [20]. The quinoprotein dehydrogenases for methanol and ethanol have a typical basic structure with a propeller fold superbarrel made up of sheets of propeller blades stabilized by the Gly/Trp docking motifs [21, 22]. The structurally related group of quinohemoprotein,alcohol dehydrogenases (QH-ADHS) has PQQ as well as heme *c* as cofactor. In the QH-ADHs for ethanol (type I [23, 24] and type II [25, 26]) and tetrahydrofurfuryl alcohol [27], the heme *c* is attached to the C-terminal part, in polyvinyl alcohol dehydrogenases [28, 29] to the N-terminal part of the enzyme. Besides for the mechanism of electron transfer between PQQ and heme *c*, QH-ADH type II is also interesting from an applied point of view since it is the primary enzyme converting ethanol into vinegar in Acetic acid bacteria. Moreover, it has the required activity and enantioselectivity for production of enantiopure glycidol and solketal, building blocks for the synthesis of various homochiral pharmaceuticals [30, 31].

Table 2

Quinohemoprotein (PQQ/heme *c*-containing) alcohol dehydrogenases

ENZYME	ORGANISM	ENZYME	ORGANISM
Ethanol DH (I)	*Comamonas testosteroni,* *Pseudomonads*	Tetrahydrofurfuryl alcohol DH	*Ralstonia* *eutropha*
Ethanol DH (II)	*Acetobacters* and *Gluconobacters*	Polyvinyl alcohol DH	*Pseudomonas* species

Table 3

Microbial alcohol oxidoreductases

ENZYME	COENZYME	COFACTOR	SOURCE
Dehydrogenases			
NAD(P)-dependent Long-, short-chain, Zn-cont., Fe-cont.	NAD(P)		Many microbes
Deazaflavin-dependent	F420		Archae
Quinoprotein MDH, ADH.		PQQ	Gram-neg. bacteria
Quinohemoprotein ADH, PVA-DH, THF-DH		PQQ/heme *c*	Gram-neg. bacteria
Flavoprotein PEG-DH, choline-DH, octanol-DH, etc.		FAD	Gram-neg. bacteria
Nicotinoprotein MNO, MDH, np-ADH		NAD(P)	Bacteria, Archae
Oxidases			
Flavoprotein MOX, AOX		FAD	Yeasts, fungi

3. GLUCOSE OXIDATION

3.1. Introduction

Phosphorylative glucose degradation is well known but many organisms are also able to oxidize glucose (and other hexose and pentose sugars) directly without phosphorylation. Direct oxidation of sugars can take place at nearly every position in the molecule. To keep the review within the limits, only enzymes are considered who oxidize glucose at the 1-position. The oxidation product in that case, the aldonolactone, becomes either spontaneously or enzymatically hydrolysed to aldonic acid. The acid can either be metabolized or, as occurs by some fungi and bacteria, excreted. In the case of wood rot fungi, it has been suggested that sugar oxidation proceeds *via* an oxidase since the H_2O_2 produced can be used for lignin degradation by peroxidases. Similarly, in the case of hexose oxidase from red seaweed species it has been suggested that the enzyme protects the organisms by producing the antimicrobial H_2O_2, more or less analogous to the function ascribed to peroxisomal oxidases.

3.2. NAD(P)-dependent glucose dehydrogenases

NAD(P)-dependent glucose (and for other sugars) dehydrogenases have been found in several organisms, including Archae [32].

3.3. Nicotinoprotein glucose dehydrogenases

The bacterium *Zymomonas mobilis* contains a nicotinoprotein glucose dehydrogenase that is active in the following interconversion, suggested to be crucial for osmotolerance of this organism [33]:

$$Glucose + E{\sim}NADP \rightarrow Gluconolactone + E{\sim}NADPH$$

$$E{\sim}NADPH + Fructose \rightarrow Sorbitol + E{\sim}NADP$$

Sugar epimerization or isomerization can in fact be regarded as an interconversion of two groups in the molecule itself. Not surprising, therefore, several of these enzymes are nicotinoproteins.

3.4. Flavoprotein glucose dehydrogenases and oxidases

Flavoprotein glucose dehydrogenase was originally found in some fungi already several years ago [34] but the finding has meanwhile not been confirmed in the literature (however, since we found dye-linked glucose dehydrogenase activity in several fungi (B.W. Groen and J.A. Duine, unpublished results), the enzyme may indeed exist). Since the fungi in which it was found also produced glucose oxidase (GOX), it would be interesting to know whether any relationship exists. On the other hand, the amino acid sequence of flavoprotein glucose dehydrogenase from *Drosophila melanogaster* shows only partial similarity with that of glucose oxidase [35].

GOX is produced by several fungi and is found in the culture media as well as in the peroxisomes [36]. Since the enzyme is commercially available and there is a big market for devices for glucose determination, a tremendous number of articles can be found dealing with the design of biosensors and test strips in which this enzyme is used. Curiously, improvement of electron transfer from the enzyme to an electrode can be achieved by coupling of PQQ to FAD or to the electrode surface [37, 38].

In contrast to GOX, as is in the name hexose oxidase from red seaweeds has far broader substrate specificity. It has a covalently bound flavin as cofactor [39] and its amino acid sequence shows insignificant similarity with that of GOX [40]. Related to its broad substrate specificity and the more or less "GRAS" status of its source, it is promising for applications in sugar-containing food where *in situ* generation of H_2O_2 is required (e.g. in dough making).

3.5.Quinoprotein glucose dehydrogenases

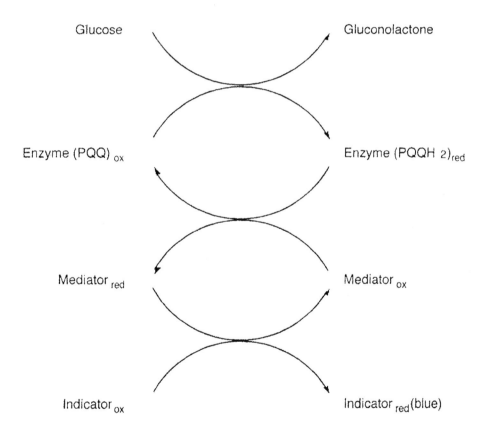

Fig. 3. Principle of glucose determination in blood *via* test strips with sGDH as sensor. Amperometric biosensor, test strip, diagnostic device, glucose monitoring, sGDH

Two quite different PQQ-containing glucose dehydrogenases occur in all strains of the bacterium *Acinetobacter calcoaceticus:* mGDH, membrane-bound glucose dehydrogenase, and sGDH, soluble glucose dehydrogenase. mGDH has been found in many Gram-negative bacteria, also in *Gluconobacter industrius* strains that are used for the industrial production of gluconic and keto-gluconic acids. The enzyme of *Pantoea citrea* is involved in pink disease of pineapple [41]. mGDH also plays a role in certain soil bacteria to liberate phosphate from insoluble calcium phosphate [42] (by the produced gluconic acid) and in nitrogen fixation [43].

Despite several attempts to demonstrate the physiological activity of sGDH in *A. calcoaceticus,* the results were negative, i.e. no active coupling of sGDH to the respiratory chain in the cell was found. On the other hand, nowadays the enzyme is used on a worldwide base in diagnostic devices (test strips (Fig. 3), amperometric biosensors) for glucose monitoring, replacing GOX for that purpose [44]. Important reasons for this are: by definition, sGDH does not react with O_2 whereas GOX does (a variable O_2 concentration in samples will affect the outcome in the case of GOX-based monitors); sGDH has an extremely high turnover number (short measuring times, small samples of blood); sGDH has a broad specificity with respect to artificial electron acceptors (enlarging the possibilities with respect to finding the most suited one).

Although the amino acid sequence of sGDH shows no similarity with those of other PQQ-enzymes [45], including that of mGDH, its 3-D structure has [46]. Also amino acid residues involved in PQQ binding have been conserved. Structural and kinetic data have provided the first evidence for the chemical mechanism of a PQQ-enzyme, showing that it concerns Ca^{2+}-assisted hydride transfer with tautomerization of C5-reduced PQQ to $PQQH_2$ as the rate-determining step (Fig. 4) [47]. The PQQ-activating Ca^{2+} enhances the rate of glucose oxidation enormously, probably by polarizing the C5-carbonyl of PQQ, in this manner assisting hydride transfer. The stabilizing effect on the semiquinone form of PQQ [48] may be relevant for oxidation of the reduced enzyme with one-electron acceptors.

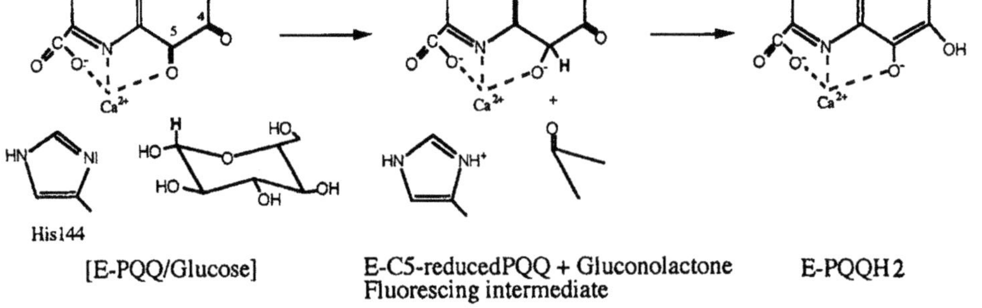

[E-PQQ/Glucose]   E-C5-reducedPQQ + Gluconolactone   E-PQQH2
Fluorescing intermediate

Fig. 4. The chemical mechanism of sGDH in the oxidation of b-D-Glucose.

4. DISCUSSION

4.1.Conclusions

For twenty years ago, the enzymology of biological oxidation was rather simple. It concerned two types of oxidoreductases, the NAD(P)-dependent dehydrogenases and the flavoprotein dehydrogenases/oxidases. Both types have distinct characteristics determined by the chemical nature of their coenzyme/cofactor: NAD-dependent dehydrogenases catalyse two-electron (hydride), flavoproteins two- as well as one-electron transfer steps. NADH is rather stable in the presence of O_2 but depending on the type of active site in which they occur, reduced flavins become easily oxidized with it. Dehydrogenases are instrumental in cases where energy generation is required. On the other hand, oxidases are probably preferred in cases where oxidation should be carried out outside the cell (the reaction or the products are too dangerous for the cytosol, or transport forms a problem), where generation of hydrogen peroxide is required, or where useful energy generation is not crucial.

In the past twenty years, it has become clear that besides the nucleotide-derived cofactors NAD and FAD, also oxidized amino acid-derived cofactors exist. This not only concerns the quinone cofactors but e.g. also the free radical forms of specific Tyr and Trp residues in the protein part of certain enzymes (e.g. in ribonucleotide reductase [49]) or of a specific cysteine residue (e.g. in galactose oxidase [50], nitrile hydratase [51], quinohemoprotein amine dehydrogenase [54, 55]). All these cofactors, except PQQ, form part of the protein chain. However, also flavins are either covalently of non-covalently bound to the protein and so far the reason of this difference is not clear. NAD(P) and F420 are the only redox-coenzymes found for dehydrogenases. All quinoprotein dehydrogenases occur in the periplasm of Gram-negative bacteria so that transport across the cytoplasmic membrane for the substrates can be circumvented in bacteria having this type of enzyme. However, the same localization has been reported for several flavoprotein dehydrogenases.

In view of the similarity in primary structures, nicotinoprotein and NAD(P)- dependent ADHs may have developed from each other or from a common ancestor. With the possession of a nicotinoprotein ADH coupled to the respiratory chain, it is possible to disconnect the reaction catalyzed from the effect of the NAD/NADH balance in the cell. In cases where the nicotinoprotein is not coupled to the respiratory chain, such a type of enzyme may be very useful for rapid interconversion of aldehydes and alcohols, and sugars and polyols.

The overview shows that quite different enzymes nevertheless can catalyze the same reaction or even the oxidation of the same substrate. The enzymes differ from each other with respect to identity of the cofactor and the primary as well as the tertiary structure. However, sGDH oxidizes glucose *via* direct hydride transfer and enzymes with TTQ, TPQ or CTQ (cysteinyl tryptophanquinone [54, 55]) as cofactor oxidize amines *via* a mechanism similar to that of enzymes using pyridoxal phosphate as coenzyme. Therefore, mechanistic diversity seems absent in enzyme types differing in cofactor and protein structure.

4.2.Diversity of catalysts for the same reaction: relics from the past or optimal tools for not fully understood tasks?

Substrate specificity overlap between enzymes belonging to the groups of NAD(P)-dependent dehydrogenases and flavoprotein oxidoreductases has already been known for some time. This was so far acceptable from the feeling that functioning in a certain organism or environment requires diversification in catalysts for a certain reaction, being in line with the common explanations for biological diversity in general (caused by adaptation of an organism to the particular niche occupied). However, from developments in the past twenty years, the correctness of this feeling can be questioned now.

The discovery of quinoprotein oxidoreductases as a third group of oxidative enzymes initially gave the impression that the possession of such exotic cofactors might confer special catalytic capabilities to these enzymes so that the elucidation of their mechanisms might unveil reasons for diversity in catalysts. However, as concluded above, this is not he case. Therefore, can a more detailed comparison between the groups reveal a reason?

When coupled to the respiratory chain, both quinoprotein and flavoprotein dehydrogenases catalyse irreversible substrate oxidation. In principle, both can carry out two- as well as one-electron transfer steps. As concluded above, the enzyme types do not differ with respect to localization and mode of binding of the cofactor. A distinct property is the redox potential, quinone cofactors having a much higher one than flavins. However, this is a thermodynamic, not a kinetic difference so that the relevance, of what seems to be a thermodynamic overkill, is not obvious for catalytic competence. In view of all these similarities, does diversity in catalysts has any meaning?

The distinction between the three groups with respect to cofactor/coenzyme identity as well as primary and tertiary protein structure, indicates that they originated independently in evolutionary history. It could be reasoned that as so many remnants from the past, they just remained in the organism in which they originated, the more as they do the job for which they originally were developed so that there is no need for it to replace them*. This idea suggests that the genes for these enzymes are more or less silent. However, as illustrated by the examples in Table 1 and 2, it concerns "vital" genes because they easily adapted to changing substrate supply during the past 50 years, yielding enzyme variants oxidizing the xenobiotics tetrahydrofurfuryl alcohol, 4-nitrobenzyl alcohol and the large PVA and PEG polymers. Moreover, this view violates the biological law of selection pressure, leading to removal of suboptimal variants in the long term. Rapidity and economy (in bioenergetic terms) in catalysis of the steps of metabolic pathways are important assets in the struggle for survival of microbes. From the kinetic parameter values and coupling site at the respiratory chain, it is clear that members of the three groups oxidizing the same substrate, are indeed different in these respects from each other (although this seems not connected to the possession of a specific cofactor). Since lateral gene transfer between microbes has so frequently been observed with respect to antibiotic resistance or xenobiotics metabolism, one would expect that if one of the three would be superior, it should have replaced the other two already long ago.

The fact that all three still exist, sometimes even in the same organism, brings us to the

---

*The latter seems irrelevant since it applies even to cases where the enzyme is not functional, e.g. because the organism does not produce the cofactor for the enzyme. This phenomenon has been frequently observed for quinoproteins where the organism produces the apo-enzyme but not PQQ (e.g. for mGDH and *E. coli*). Nevertheless, although these bacteria are kept in Culture Collections for many years and it is unlikely that any PQQ is present in this environment, the apo-enzymes are still produced and have normal turnover numbers when reconstituted with PQQ).

opposite explanation, i.e. that the diversity has a meaning. Striking examples of what seems illogical use of enzyme type in the first instance can be derived from the overview. At least six gene products are involved in the biosynthesis of PQQ. Nevertheless, Pseudomonads do this effort to produce it only for PQQ-containing ADH whereas they always synthesize the multipurpose cofactor FAD that is used by much more dehydrogenases (including FAD-containing ADH in these bacteria) *. Probably related to the impossibility of carrying out 1-electron transfer steps, an at least partially new, probably exclusive, respiratory chain is necessary as electron acceptor for nicotinoprotein MNO in *A. methanolica*. Why is this reaction not catalysed by an NAD-dependent methanol dehydrogenase, for which a common respiratory chain can be used? Glutathione is the coenzyme for NAD-dependent formaldehyde dehydrogenase in many organisms, from human beings to bacteria. However, although the amino acid sequence is rather similar, Actinomycetes use mycothiol as coenzyme for their formaldehyde dehydrogenase. When *Pscudomonas aeruginosa* grows on ethanol, as revealed by the suicide inhibitor cyclopropanol, it contains an active quinoprotein as well as an NAD-dependent ADH [52]. Depending on the strain, *Gluconobacter oxydans* has a sorbitol dehydrogenase with FAD, PQQ or PQQ/heme c as cofactor [20]. Similarly, one strain of this bacterium has been found to contain an L-sorbose dehydrogenase with PQQ, another with FAD as cofactor [11, 53]. For the examples given here, no data are available that can really explain why a certain type of catalyst is preferred over the others, but assuming that it was a deliberate choice in evolutionary history, it must concern subtleties that are not known since they may have been overlooked.

The exclusivity, illustrated above with a few examples, in type of enzyme use for a certain reaction is widespread. Although the latter *suggests* that diversity should have significance, it does not *prove* it. Moreover, since the comparison of enzymatic properties for the three groups (see above) did not reveal any clue, this could be interpreted in disfavour of the suggestion. However, the negative result may also reflect our ignorance, related to the adagio "the more we know the less we understand". On the other hand, when giving it a more positive interpretation, our ignorance could also be a sign that so far enzymology has focused on aspects that are irrelevant for explaining the diversity. If so, enzymes should be studied in the context of their functioning. Restricting this to microbial oxidations, microbial physiology and ecology should provide clues for this context so that the specific *in vivo* situation can be mimicked and the various oxidoreductases for a certain reaction can be compared. This may reveal why diversity exists, either because biological variation requires diversity at the molecular level or because selection pressure is unable to remove the functionally less perfect enzyme types.

Acknowledgments. The contribution of dr. A. Tani in preparing the figures is highly appreciated.

REFERENCES

1. J. Vonck, N. Arfman, G.E. de Vries, J. van Beeumen, E.E.J. van Bruggen and L. Dijkhuizen, J. Biol. Chem., 266 (1991) 3949.

---

*Since approximately 200 different flavoproteins are known but less than 20 quinoproteins (with PQQ as cofactor), the discrepancy between effort and benefit applies in general.

2. M. Misset-Smits, P.W. van Ophem, S. Sakuda and J.A. Duine, FEBS Lett., 409 (1997) 221.

3. A. Norin, P.W. van Ophem, S.R. Piersma, B. Persson, J.A. Duine and H. Jornvall, Eur. J. Biochem., 248 (1997) 282.

4. A.R. Klein, H. Berk, E. Purwantini and R.K. Thauer, Eur. J. Biochem., 239 (1996) 93.

5. L.V. Bystrykh, N.I. Govorukhina, L. Dijkhuizen and J.A. Duine, Eur. J. Biochem., 247 (1997) 280.

6. N. Kato, T. Yamagami, M. Shimao ans C. Sakazawa, Eur. J. Biochem. 156 (1986) 59.

7. P. Schenkels and J.A. Duine, Microbiology, 146 (2000) 775.

8. M. Sugimoto, M. Tanabe, M. Hataya, S. Enokibara, J.A. Duine and F. Kawai. J. Bacteriol.,183 (2001) 6694.

9. J.B. van Beilen, G. Eggink, H. Enequist, R. Bos and B. Witholt, Mol. Microbiol., 6 (1992) 3121.

10. K.D. James, M.A. Hughes and P.A. Williams, J. Bacteriol., 182 (2000) 3136.

11. Y. Saito, Y. Ishii, H. Hayashi, Y. Imao, T. Akashi, K. Yoshikawa, Y. Noguchi, S. Soeda, M.Yoshida, M. Niwa, J. Hosoda and K. Shimomura, Appl. Environ. Microbiol., 63 (1997) 454.

12. E. Shinagawa, K. Matsushita, 0. Adachi and M. Ameyama, Agr. Biol. Chem., 46 (1982) 135.

13. J.R. Woodward, in: G.A. Codd, L. Dijkhuizen, F.R. Tabita (eds.), Advances in Autotrophic Microbiology and One Carbon Metabolism, Kluwer Academic Publishers, Dordrecht 1990, p. 193.

14. E. Varela, M.J. Martinez and A.T. Martinez, Biochim. Biophys. Acta, 1481 (2000) 202.

15. M.W. Fraaije, W.J.H. van Berkel, J.A.E. Benen, J. Visser and A. Mattevi, Tr. Biochem. Sci., 6 (1998) 206.

16. J.A. Duine, J. Biosci. Bioeng., 88 (1999) 231.

17. J.M.J. Schrover, J. Frank, J.E. van Wielink, J.A. Duine, Biochem. J., 290 (1993) 123.

18. A. Diehl, F. von Wintzingerode and H Goerisch, Eur. J. Biochem. 257 (1998) 409.

19. M.A.G. van Kleef and J.A. Duine, Arch. Microbiol., 150 (1988) 32.

20. 0. Adachi, Y. Fujii, Y. Ano, D. Moonmangmee, H. Toyama, E. Shinagawa, G. Theeragool, N. Lotong and K. Matsushita, Biosci. Biotechnol. Biochem. 65 (2001) 115.

21. Z. Xia, W. Dai, Y. Zhang, S.A. White, G.D. Boyd and F.S. Mathews, J. Mol. Biol., 259 (1996) 480.

22. T. Keitel, A. Diehl, T. Knaute, J.J. Stezowski, W. Hoehne and H. Goerisch, J. Mol.Biol., 297 (2000) 961.

23. G.A.H. de Jong, A. Geerlof, J. Stoorvogel, J.A. Jongjan, S. de Vries and J.A. Duine, Eur. J. Biochem., 230 (1995) 899.

24. H. Toyama, A.Fujii, K. Matsushita, E. Shinagawa, M. Ameyama and 0. Adachi, J. Bacteriol., 177 (1995) 2442.

25. H. Takemura, K. Kondo, S. Horinouchi and T. Beppu, J. Bacteriol. 175 (1993) 6857.

26. K. Kondo and S. Horinouchi, Appl. Environ. Microbiol. 63 (1997)1131.

27. G. Zarnt, T. Schroeder and J.R. Andreesen, Appl. Environ. Microbiol. 63 (1997) 4891.

28. M. Shimao, T. Tamogami, S. Kishida and S. Harayama, Microbiology, 146 (2000) 649.

29. N. Takizawa, Data base Entry D83772.

30. U. Wandel, S. Salgueiro Machado, J.A. Jongejan and J.A. Duine,

Enzyme Microb. Technol., 28 (2001) 233.

31. A. Geerlof, J.B.A. van Tol, J.A. Jongejan and J.A. Duine, Biosci. Biotech. Biochem., 58 (1994) 1028.

32. M.-J. Bonete, C. Pire, F.I.L. Lorca and M.L. Camacho, FEBS Lett., 383 (1996) 227.

33. H. Loos, R. Kramer, H. Sahm and G.A. Sprenger, J. Bacteriol. 176 (1994) 7688.

34. T.-G. Bak, Biochim. Biophys. Acta, 139 (1967) 277.

35. B.L. Keplinger, X. Guo, J. Quine, Y. Feng and D.R. Cavener, Genetics 157 (2001) 699.

36. M. Kiess, H.-J. Hecht, and H.M. Kalisz, Eur. J. Biochem., 252 (1998) 90.

37. E. Katz, V. Heleg-Shabtai, B. Willner, I. Willner and A. Bueckmann, Bioelectroch. Bioenerg. 42 (1997) 95.

38. H. Shinohara, T. Kusaka, E. Yokota, R. Monden and M. Sisido, Sensors and Actuators, 65 (2000) 144.

39. B.W. Groen, S. de Vries and J.A. Duine, Eur. J. Biochem. 244 (1997) 858.

40. 0. C. Hansen and P. Stougaard, J. Biol. Chem. 272 (1997) 11581.

41. C.J. Pujol and C.l. Kado, Microbiology, 145 (1999) 1217.

42. A.H. Goldstein, K. Braverman and N. Osorio, FEMS Microbiol. Lett., 30 (1999) 295.

43. M. Flores-Encarnacion, M. Contreras-Zentelia, S-. Soto-Urzua, G.R. Aguilar, B.E. Baca and J.E. Escamilia, J. Bacteriol., 181 (1999) 6987.

44. N. Kaufman, Glucotrend:Evaluation of a new system for determining blood glucose, Roche Diagnostics 1998.

45. A.-M. Cleton-Jansen, N. Goosen, T.J. Wenzel and P. van de Putte, J. Bacteriol. 170 (1988) 2121.

46. A. Oubrie, H.J. Rozeboom, K.H. Kalk, J.A. Duine and B.W. Dijkstra, J. Mol. Biol. 289 (1999) 319.

47. A.R. Dewanti and J.A. Duine, Biochemistry, 39 (2000) 9384.

48. A. Sato, K. Takagi, K. Kano, N. Kato, J.A. Duine and T. Ikeda. Biochem. J., 357 (2001) 393.

49. J.-A. Stubbe, J. Ge and C.S. Yee, Tr. Biochem. Sci. 26 (2001) 93.

50. M.M.Whittaker, C.A. Ekberg, J. Peterson, M.S. Sendova, EP. Day, J.W. Whittaker, J. Mol. Catal. B: Enzymatic, 8 (2000) 3.

51. M. Tsujimura, N. Dohmae, M. Odaka, M. Chijimatsu, K. Takio, M. Yohda, M. Hoshino, S. Nagashima and I. Endo, J. Biol. Chem. 272 (1997) 29454.

52. A. Groeneveld, M. Dijkstra and J.A. Duine, FEMS Microbiol. Lett. 25 (1984) 311.

53. A. Asakura and T. Hoshino, Biosci. Biotechnol. Biochem. 63 (1999) 46.

54. I. Vandenberghe, J.-K. Kim, B. Devreese, A. Hacisalihoglu, H. Iwabuki, T. Okajima, S. Kuroda, O. Adachi, J.A. Jongejan, J.A. Duine, K. Tanizawa and J. Van Beeumen. J. Biol. Chem., 276 (2001) 42923.

55. A. Satoh, J.-K. Kim, B. Devreese, I. Vandenberghe, A. Hacisalihoglu, I. Miyahara, T. Okajima, S. Kuroda, O. Adachi, J.A. Duine, J. Van Beeumen, K. Tanizawa and K. Hirotsu. J. Biol. Chem., 10.1074/jbc.M109090200.

Index of Authors

Index of Key words

Printed and bound by CPI Group (UK) Ltd, Croydon, CR0 4YY

08/05/2025

01865007-0006